Experiments for Young Botanists

C. T. Prime

London: G. Bell & Sons Ltd

PUBLISHED BY G. BELL & SONS LTD,
YORK HOUSE, 6 PORTUGAL STREET, LONDON, WC2

PRINTED IN GREAT BRITAIN BY
THE ANCHOR PRESS LTD
TIPTREE, ESSEX

ISBN 0 7135 1667 4

For the wonderful and secret operations of Nature are so involved and intricate, so far out of reach of our senses, as they present themselves to us in their natural order, that it is impossible for the most sagacious and penetrating genius to pry into them, unless he will be at the pains of analysing Nature, by a numerous and regular series of Experiments; which are the only solid foundation whence we may reasonably expect to make any advance, in the real knowledge of the nature of things.

Stephen Hales, *Vegetable Staticks*, 1727

Contents

Plates

Preface

PLANTS attract people because of their beauty and because of the many other needs of man they supply. So many, perhaps even the majority, get great pleasure out of trying to grow plants, and in this way taking a small part in creating the beauty that the plant world displays. Part of this book is therefore concerned with growing plants, and finding out what they need to grow well. It is also concerned with looking at plants in order to see a little of the detail that the casual glance would pass over, and so a number of things are suggested that the young botanist or gardener can do to gain an insight into plant structure and behaviour. Moreover, most people like to know a little of how things work, and most of the book is taken up with a number of experiments, some simple and easy, others rather more complicated, which shed a little light on the complexities of the plant's machinery. There is almost an infinity of experiments to choose from, but all of these in the book can be carried out with virtually no apparatus, or with the equipment available in a chemistry set. The chemicals are, on the whole, cheap and fairly easy to obtain from the various supply agencies. Alcohol is in some ways the most difficult, since duty-free spirit is not on sale and coloured methylated spirit will not do for experiments where colour changes have to be looked for. In most cases these can be replaced by isopropyl alcohol, or by methylated spirit decolourised with the aid of a little animal charcoal.

In biological work the understanding of the significance of the results is often most important, and this is the reason for the experiments being accompanied by or almost embedded in the text. All the same, theory has been cut to a minimum, and Latin names only given where it is necessary to make clear the species intended.

I should like to acknowledge and thank the following for permission to reproduce the following photographs and drawings: Glasscrops Crops Research Institute, Rustington, Sussex, for plate 5, and Plant Protection Ltd for plates 11 and 12. I should also like to express my thanks to Messrs Bell for the help and encouragement they have given me in the prepaiation of this book.

C. T. PRIME

Farleigh,
Warlingham, Surrey
January 1971

1 Growing seeds in soil

SOIL can vary in many different ways. Take some soil from your garden and examine it. It may be heavy or light in weight, stony or sandy, wet or dry, clayey or chalky, dark or light in colour, easy to dig or hard to turn over. All the same, most soils contain a proportion of larger mineral particles, often sandy in character, and successive proportions of particles of smaller size, the smallest being of clay. Sandy soils have a comparatively greater amount of large particles, while clay soils have a larger percentage of particles of very small size. The relative amounts of these particles affect the air and water content of the soil. In general, the larger the particles, the greater the air content, for there is a greater space between the particles; the smaller the particles, the smaller the spaces, and so water is held by capillarity and does not drain away so easily. All this is well known, for everybody knows that clay soils are usually wet and sticky while sandy soils are dry. It does not necessarily follow that this is always so, for sands can lie in hollows of clay and in such circumstances they can be very wet.

Material formed from the remains of dead plants and animals is also incorporated in the soil. This is known as humus and gives the soil its dark colour. For example, peaty soils are particularly rich in organic matter, for under the right conditions the peat breaks down to form humus. These soils are also often wet, for peat retains water well. Further, soil is populated by a host of living organisms ranging from quite large animals down to microscopic plants and animals. The microscopic life is so numerous that it greatly affects the fertility of the soil. Relatively large animals like worms make burrows and turn the soil over, but the smaller organisms bring about chemical changes which increase the amount of available plant food in the soil.

To prepare some surface soil for small experiments it is best to turn it over with a spade and then to rub it through a garden sieve to remove the large stones. To see how well plants grow in it take a number of plant pots or boxes all of the same size and scrub them clean. Place a crock (a broken piece of pot) over the holes at the base of the pot and almost fill it with soil. Take a packet of seed, say mustard, and sow a number, e.g. ten, taking care to space them evenly. If you are going to try a number of experiments with the same size of pot it is worth while making a little circular grid of wood which just fits into the top of the pot, and in which you have drilled ten evenly spaced holes. Then you just spread the seeds over the surface so that one drops into each hole and sweep the remainder back into the packet. Remove the grid,

Table 1   *The fresh weight of radish plants*
from germination to the age of fourteen weeks

Number of weeks after germination	Fresh weight (gms)
—	0·01
1	0·07
2	0·30
3	0·45
4	2·2
5	9·2
6	28·0
7	44·0
10	59·1
13	62·5
14	65·3

cover the seeds with soil to a depth that corresponds to the size of the seed, press the soil down firmly and leave them in a warm place. Water them when necessary, and as the seedlings appear make a note of the date and the number. Measure the height of plants every two or three days, count the number of leaves, take note of their colour and record the time of flowering. In this way you will get a complete record of the growth of the mustard plants. It may be necessary to reduce the number of seedlings in the pot if it becomes too crowded. Table 1 shows a result obtained by growing a large number of radish plants and taking a number each week for measurement.

In all the experiments described in these chapters it is highly desirable to duplicate them, or, better still, to set up a number of each. For living organisms, even mustard seeds, are variable and so is the soil, the other materials, and the conditions you are working with. You minimise the effects of these variables by this replication and by judging your results by a strict comparison of your two sets of experiments. It is worth a lot of trouble to see that your experiments are carefully prepared in all respects, for only then can you expect to obtain trustworthy results.

By growing pots of mustard seeds in different conditions you will soon find out some of the essential requirements for plant growth. You can place pots sown with seeds in a refrigerator when you will find no growth at all. Also, you can repeat the experiment with three sets of pots, giving the first hardly any water, the second a little and the third plenty. It is more scientific to give a measured amount each day, but it is rather hard to specify exact quantities, since so much depends on the size of the pots and boxes you may use. In this way you can easily verify what is common knowledge to all, namely that plants require water, a suitable temperature and some substances from the soil in order to grow well.

It is interesting and not too difficult to study further the growth of plants in soil. At the beginning of the chapter, soil was said to contain different kinds of particles. To show some of the effects of these, you can grow mustard seed in pots containing clay, in pots containing sand, and in pots containing as many mixtures of sand and clay as you choose to make. Clay can be obtained from a brickyard, but modelling clay from a shop will do equally well, while sand can be obtained from a builders' yard. Peat is also readily obtained from a nursery, and you can try to grow mustard in a mixture of clay, sand and peat, and in this way obtain a mixture which is more like a true soil, and in which the plants will grow better.

You will now have found that a plant like mustard grows better in ordinary garden top soil than in the soil constituents alone. But you will find that there is not much plant food in the top soil. Try growing a second crop of mustard after the first in the same soil. To do this, break up the soil and the plant roots, replace them in the pots and sow fresh sets of seeds exactly as before. You will find that the mustard does not grow as well, and hence you may infer that a part of the plant food has been taken up by the previous crop and removed with it. If you have some other pots that have been used for growing mustard you can try growing another kind of seed, say cress, in the same soil. You will not see quite the decline in vigour as in the pots where you have grown the same plant twice on the run. This is because although plants require the same foodstuffs, they require different amounts of the chemicals concerned. For example, one plant may require relatively large amounts of phosphate, whereas another may need very little. In this simple experiment you see one of the reasons why farmers rotate their crops, i.e. why they do not grow the same crop year after year in the same soil.

You can try further experiments with the spent soil left over from growing mustard. Make it into a heap, turn it over thoroughly and divide it into two parts. Spread one part out to dry and then keep it in a dry place. Keep the other half moist for a month and then once more make up the two sets of pots with mustard seed as before. Treat them in exactly the same manner and in due time you will find that the soil that has been kept moist has recovered some of its fertility and that it now grows a good crop once more. The dry soil does not do so, and it is clear that a rejuvenating process has gone on in the soil that has been left to itself in moist conditions.

More can be found out about this rejuvenating process by taking some plant remains like grass mowings or dead leaves, chopping them up and making a very fine mixture of them. Equal parts of the mixture are now added to clay, sand and top soil and once more the pots are sown with mustard, together with another series from which the plant mixture is missing. The latter serve as controls. You will find, as you might expect, that the addition of plant remains improves the growth of the mustard, and furthermore the effect is more pronounced in the top soil. If you tip out the pots carefully, and examine them for the remains of the grass mixture you added, you will find more of the remains are visible in the sand and clay than in the top soil. In other words, the plant remains become incorporated in the top soil more quickly than in the others, and, further, it would not be unreasonable

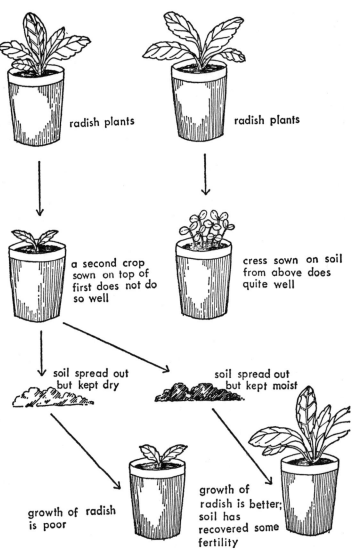

radish plants

radish plants

a second crop sown on top of first does not do so well

cress sown on soil from above does quite well

soil spread out but kept dry

soil spread out but kept moist

growth of radish is poor

growth of radish is better; soil has recovered some fertility

Fig. 1   Experiments on soil fertility

to connect this observation with the increase in the fertility. The plant remains are therefore a source of plant food, but they have to undergo some change before they are usable, and this occurs more quickly in top soil than in sand or clay, and more quickly in wet than in dry soil (Fig. 1).

You will now have realised that the number of experiments of this kind is infinite, and no doubt some will suggest themselves to you. Try repeating the experiments, for example, adding a small amount of soot, or a small amount of chalk or any one of the fertilisers sold at the horticultural shops.

From many more complex experiments, but, all the same, very similar experiments to those you have done, scientists have found that plants require water and certain mineral elements from the soil to grow. You may then try to grow mustard seeds in water and the mineral salts alone. This can be done and the easiest way is to make up the following mixture of salts and dissolve about a tablespoonful in a gallon of water:

> 5 gm potassium chloride
> 5 gm potassium phosphate (KH_2PO_4)
> 5 gm magnesium sulphate ($MgSO_4$ $7H_2O$)
> 20 gm calcium nitrate ($Ca(NO_3)_2$ $4H_2O$)
> 2 gm ferric chloride ($FeCl_3$)

Alternatively a suitable mixture can be purchased from a firm dealing in chemicals and biological supplies* The solution should be used to water the seeds sown in sand. To make sure the sand is free of any soluble plant food, you must wash it thoroughly with water first. Take a pail or bowl with holes in the bottom, block them partly with rag and then half fill with sand. Water can then be allowed to drip through until it runs quite clean. Freshly bought sand is virtually free of plant nutrients, but it is as well to wash it to make sure. Alternatively, a material like vermiculite can be used in which to grow the plants. Care is required to make the plants grow well, for they can easily succumb to drought or disease if not well looked after.

* See Appendix 1.

As a result of experiments of the kind described in this chapter, soil mixtures known as John Innes composts have been standardised as being most suitable for the germination and growth of seeds. They are so called because they were first devised by the John Innes Horticultural Institute. These composts can be bought from horticultural shops, but you can make them up for yourself if you are prepared to take the trouble. The basis of the John Innes compost is a mixture of two parts of loam, one part of sand and one of peat. Loam is the most important constituent to make properly, and it should be done by cutting turf and stacking it in piles (it must be damp, why?) for about six months until it is completely rotted down. Perhaps for your purpose any good soil that you can find might be used. The loam should be sieved and mixed with the sand. Peat can be purchased and this too should be rubbed through the sieve. The composts sold at the shop have been sterilised to kill the spores of pests and diseases in the soil, but this is difficult to do with any large quantity of soil at home. Basically the process consists of heating the soil to about the boiling point of water or above. You can try small quantities in a pressure cooker if you have one, or even in an oven. If you do try, the loam must be dry to begin with and not wet. You need only sterilise the loam, and although you can sterilise the sand and loam together, you cannot sterilise the whole compost mixture, as chemical changes go on which make it much less suitable for plant growth. For instance, the amount of nitrogen available can be increased to a level too high for good seedling growth.

To complete the compost it is necessary to add one and half ounces of superphosphate and three-quarters of an ounce of chalk to every bushel of soil; two or three pailfuls may be taken as equal to a bushel. The compost will be found suitable for growing the seeds of many common garden vegetable plants. You can compare the growth of mustard in J.I. compost with that in ordinary soil.

For growing the seedlings on into individual plants a slightly different compost is used. This is called the J.I. potting compost and this contains loam, peat and sand in the ratio $7:3:2$, that is rather more loam and slightly less sand that the seed compost.

To this is added the same amount of chalk ($\frac{3}{4}$ oz to the bushel) and a quarter of a pound of what is called J.I. base. Made up by weight it consists of

> 2 parts of hoof and horn (a source of nitrogen)
> 2 parts of superphosphate
> 1 part of sulphate of potash

and it can be bought from many horticultural suppliers. Again, growth of mustard in this compost can be compared with that in the standard J.I. seed compost.

More recently, partly owing to the difficulty of obtaining loam of suitable quality, other composts have been devised. A loamless compost suitable for sowing seeds can be made by mixing equal quantities of sphagnum peat and fine sand. By itself this mixture is deficient in essential nutrients so it is necessary to add the following

> 1 oz superphosphate
> $\frac{1}{2}$ oz potassium nitrate
> 3 oz ground chalk

to about every two or three pailfuls of the mixture. It should be thoroughly mixed by turning over many times with a spade. For growing on seedlings to maturity, additional reinforcement of the basic substances is required and the following formula is suggested. Use three-quarters by volume of the peat and a quarter by volume of the fine sand and add to this mixture

> $\frac{1}{2}$ oz ammonium nitrate
> 1 oz potassium nitrate
> 2 oz superphosphate
> 3 oz ground chalk

There may still be slight deficiencies of what are called the trace elements and probably the easiest way to get over this difficulty, where only small quantities of the compost are being made, is to add a little of a liquid fertiliser that can be bought to meet this need.

With these composts you can go on to try to raise the seedlings of as many kinds of plants as you choose. With the majority of the seed packets you can buy at the shop you will probably have little difficulty, but it is always interesting to sow a definite number and record the percentage that germinate. You may also be inclined to try the seeds of some wild plants, in which case you may meet some different patterns of germination behaviour. Some seeds require very special conditions for germination which may be referred to in later chapters. You can also try to germinate some common seeds of plants seen in the greengrocer's and grocer's shops, for instance walnuts and other nuts, date stones, orange pips, but you may well find that some of them require a really hot place like the top of a radiator to get them started.

2 More about plant nutrition

YOU will have gathered from the first chapter that plants not only require certain mineral nutrients to grow but also that they require them in different amounts and proportions. Some, like many of our important crop plants, are gross feeders, while others are able to grow well with a much smaller supply of nutrients. This is one of the reasons why some wild plants are only to be found in rich soils, while others flourish in very poor soils. Try the effect of adding small quantities (a teaspoonful or less) of potash (potassium sulphate or chloride) to tomato seedlings in pots, keeping other plants from the same batch as a comparison. These latter are spoken of as controls. Try a similar experiment, adding small quantities of ammonium sulphate to cabbage seedlings grown singly in pots, again keeping some seedlings from the same batch as controls. In the second experiment you may find the cabbage plants becoming rather dark green in colour and coarse and rank in growth. These are the effects of over-manuring with nitrogen and you can often see similar symptoms in plants growing on manure heaps. Deficiency symptoms caused by too little nitrogen are also common. Stunted growth, early flowering and the presence of reddish tints in the older leaves are the chief signs to be looked for. They may be seen in plants growing in extreme conditions, as on gravel paths or the tops of walls, or among vegetables growing on an allotment on very poor soil.

The deficiency symptoms caused by lack of adequate minerals can each be recognised by an expert as easily as the diseases caused in man by a lack of adequate vitamins. Knowledge of the exact requirements of crop plants is very important, for it is so obviously the way to greater yields and more food for human consumption.

However, getting plants to grow well is not entirely a matter of having mineral nutrients in the soil. They may be present in some soils, but for some reason or another the plant may be unable to take them up. There are several reasons for this and one of the most important is concerned with soil acidity. Chalky soils are alkaline, and moorland are acid. They are very different soils and they grow very different plants, and the difference in acidity in part accounts for this. Acidity is due to the concentration of hydrogen ions (see a chemistry book for further explanation) and at a concentration of 10^{-7} the reaction is neutral. This concentration is written on what is usually called the pH scale, as $pH = 7.0$. If the concentration increases, the pH figure decreases; thus pH 6·0 means a concentration of hydrogen ions ten times as great as that of pH 7·0. Chalky soils have a pH of the order of 8·0, while boggy soils may have a pH of 4·0. pH is easily measured by the use of indicators, and you can buy special outfits for measuring the pH of soils. However, a small supply of B.D.H. soil indicator and some barium sulphate will enable you to make a determination.

Take a test tube and put in soil to the depth of $\frac{1}{2}$ in. (1 cm) or so. If available add about an equal quantity of barium sulphate (rather less for sandy soils, rather more for clayey soils). This has the property of making the particles settle quickly, and though desirable it is not absolutely necessary. Then fill the tube three-quarters full with distilled water and add some of the indicator to a depth of about 1 in. (2 cm), shake thoroughly and allow to settle. The label on the bottle gives the pH corresponding to the colour of the indicator. Acid soils turn the indicator red while alkaline soils turn it dark green.

You can try the effect of using smaller and larger quantities of soil in the tube, and you will find that it makes no difference to the result. You can even add quite a lot of distilled water and still the pH remains the same. This illustrates one property of soils, namely that they are strongly buffered and seem to resist change. There are ways of altering the acidity of soils, but it is easier to show the effects of different pH's by making use of solutions containing the minerals necessary for plant growth and then adding to them various solutions buffered like the soil itself to a constant pH. Make up the nutrient solution as mentioned on p. 15 and the phosphate solution as mentioned later in Chapter 16 on p. 102. An appropriate mixture of these two is neutral, and a solution can be made either acid or alkaline by the addition of one or other of the phosphates. You can therefore obtain the following:

(a) nutrient solution plus acid phosphate
(b) nutrient solution plus distilled water as a control
(c) nutrient solution plus the alkaline phosphate.

Check the pH's carefully with the test indicator and then proceed to grow any plant of your choice in washed clean sand as described on p. 15. You can sow three to five seeds in each pot and then reduce the number to one soon after germination, or you can grow the plants separately and transfer them to the pots of sand. Then water the plants with solutions (a), (b) and (c) and compare their growth carefully. As in previous experiments it is desirable to set up several pots with each of the three treatments.

The mineral nutrition experiments described so far have all made use of sand or vermiculite as a base. It is possible to grow the plants directly in the mineral solutions themselves, though it is essential to take certain precautions. It is essential to maintain the solutions at neutrality and it is also necessary to replace them frequently. To set up this experiment, get some large, wide-mouthed jars which will hold about two quarts and wash them thoroughly with a detergent and rinse them several times with water to make sure that they are absolutely clean. Make a dark paper wrapper that will fit round each jar and exclude the light. Otherwise the solution will turn green owing to the growth of algae. The wrapper should slip off quite readily so that you can see how root growth is proceeding. Fill the jars with the culture solution to about $\frac{1}{2}$ in (1 cm) below the top. It may be better to use the solution at half-strength until the plant is established.

Fit a cork stopper to each jar and make three holes in it. The stopper must fit tightly so as to hold the plants erect. Dip each in melted paraffin wax, drain off the surplus and allow it to set. Grow some seedlings like barley or sunflower in sand or vermiculite until they are an inch or two high. Take three sturdy ones and fit them into the holes of the cork (Fig. 2), packing around them sufficient cotton wool to hold them firmly in position. It is very important that the cotton wool should not make contact with the solution, for it will draw it up and so wet the surface of the seedling that it will rot and die.

Before finally placing the seedlings in the solution, check the acidity and shake the solution well so that it is aerated. As the seedlings take up the water, add distilled water every few days and after a week or so remove the two weaker seedlings. Continue to repeat the pH measurements, keeping the solution neutral, and replace the whole solution every week or fortnight according to the rate of growth. In this way you should be able to grow a plant to the stage of flowering and the setting of seed in a solution only. You can also do further experiments leaving out one or other mineral nutrient and seeing what happens.

There are ways of directly altering the acidity of soil. An easy way to prepare an acid soil is to omit the chalk from the John Innes compost (described on p. 17) and to add small quantities of flowers of sulphur instead. You can then have the J.I. compost by itself as a control, the J.I. minus the chalk and the J.I. minus the chalk but with the flowers of sulphur added. Compare as described before the growth of whatever plants you please in the three composts.

The effect of soil acidity can be direct or indirect. One well-known indirect effect of soil alkalinity caused by the calcium in chalky soils is that iron gets locked up in the soil and becomes unavailable to the plant. Since iron is necessary for the formation of chlorophyll, the leaves of plants in these circumstances turn a

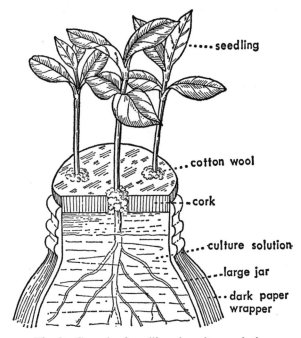

Fig. 2   Growth of seedlings in culture solution

yellow-green, a condition known as chlorosis. Careful search of plants growing on chalky soils in the wild will usually reveal leaves showing this condition. To add more iron to the soil is useless, for it just gets changed into insoluble iron phosphates or hydroxides to add to those already there. Recently chemicals have been found which get over the difficulty. They are known as chelates or sequestrenes and in these the iron is united with an organic compound that alters its properties slightly. The metallic portion is retained so that phosphates and hydroxides are not produced, and it does remain available to the plant.

The difference can be shown very clearly if, say, a little sulphate of iron is dissolved in water and a solution of lime added. A bulky brown insoluble precipitate is formed and this is what happens in limy soils. If some sequestrene iron compound (it may be purchased under this name) is treated in the same way no precipitate is formed.

The effect of this compound on the growth of plants is most pronounced in the case of so-called lime-hating or calcifuge plants like heathers and rhododendrons which so often show symptoms of chlorosis when attempts are made to grow them in limy soils. When sequestrene is added to the soil at a strength of 1 in 1000 the chlorotic symptoms disappear in a short space of time. Tomatoes and the egg plant (*Solanum capsicastrum*) respond beneficially in this way and are amongst easily and quickly grown plants which can be used for experiment. Grow batches of each in chalky soil and similar soil to which sequestrene has been added.

In general, it seems that other mineral nutrients like manganese may also be rendered unavailable at high pH levels. Not so molybdenum, of which the reverse is true. Notice also that the iron, when rendered unavailable, is turned to insoluble phosphate or hydroxide. You might expect at once that manuring with phosphate would, at high pH, tend to make iron unavailable, and such is indeed the case. Thus you begin to see some of the complexities of plant nutrition.

Let us turn to the other end of the pH scale for another indirect effect. When soils become acid, substances may be brought into solution which may benefit the plant, but there are others which may be more or less toxic to plants. Chickweed, dandelion and self-heal are quite tolerant of acidity, but the effect of acidity on a clay soil is to increase the solubility of aluminium to which these plants are sensitive. In consequence they are rare on such soils, but on acid sandy soils with little alumina they may be common. This can be shown by growing these plants as before in washed clean sand, supplying a normal nutrient solution plus a solution of aluminium sulphate, keeping as before a set of plants as controls.

Other mineral elements in the soil are more directly toxic than aluminium. Lead and other heavy metals are particularly so, and very few plants can tolerate them in any quantity. This accounts for some of the unusual plants like the spring sandwort (*Minuartia verna*) and a form of alpine penny cress (*Thlaspi alpestre*) which are to be found in the neighbourhood of lead mines. It also accounts for the appalling state of some soil heaps in mining districts on which almost nothing seems able to grow. Much research is going into the problem of reclaiming such areas and the chief hope seems to lie in finding varieties of grasses and similar plants that are tolerant of relatively large quantities of these heavy metals, and then establishing them with the help of balanced fertilisers.

Lastly, an effect of soil on flower colour is worthy of mention. The commonly grown hydrangea has flowers that are normally rose red, but when the plant is grown in acid soils they turn a deep blue. Apparently, these colours depend on the presence of aluminium in the flower buds before opening. The blue colour can be obtained by adding soluble aluminium sulphate to the soil, and, as we have seen, in the presence of acids the aluminium is released and is available to the plant. Conversely the pink colour can be retained by adding lime which tends to lock up the available aluminium in the soil. If you have any hydrangea plants you can experiment with them.

3 Different kinds of seeds and their structure

SEEDS contain a baby plant, or embryo as it can be called, together with a food store for growth. This is enclosed in a protective coat called the testa. This is a simple enough statement, but within its terms there is room for infinite variety of structure, size and pattern. A soaked pea or bean familiar to all shows one very common arrangement: an outside covering the testa, that can be easily pulled off, inside which there are two parts which readily separate. These are the seed leaves (called cotyledons) which contain the food reserve, while between them a little search will reveal a tiny shoot called the plumule and a rather larger and longer rootlet called the radicle. The seed leaves are attached to these, for only through the cotyledonary stalks can the food reserve ever reach the tiny plumule or radicle. That is

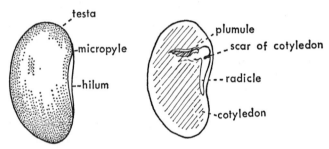

Fig. 3   Structure of bean seed

all there is, but two points must be stressed, firstly that there are two cotyledons, and secondly all the food reserve is stored within the seed leaves which are really part of the embryo (Fig. 3).

The flowering plants are divided into two large series, the dicotyledons and monocotyledons having two and one cotyledons respectively. As examples, beans and peas are dicotyledonous while maize and all the cereals are monocotyledons. Many seeds store the food reserve in a tissue separate from the embryo, in which case the cotyledons and often the whole embryo are very small, for the bulk of the seed is due to the food reserve or endosperm as it is called. The majority of the monocotyledonous seeds are endospermous, but both kinds are widely represented in the dicotyledons.

To see something of the variation in seed structure buy several packets of seeds and soak them for twenty-four hours or so. As far as possible obtain packets of large seeds, for they are obviously easier to investigate. For very small seeds it may be necessary to use a microscope. Collect or, if necessary, buy many other seeds not usually sold in packets, like apple pips and lemon pips, date and plum stones, various nuts and soak them in the same way. Then try with the aid of a penknife and some mounted needles to pull them to pieces and make out their structure. In

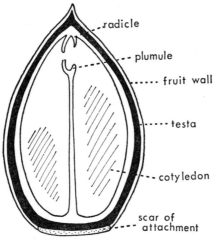

Fig. 4   Structure of the acorn

general it is better to try to prise the seeds apart rather than to cut them, for by this method they may separate readily into their component parts.

In the flowering plants the seeds are always contained in a fruit of some kind; we usually think of a fruit as being fleshy or succulent, but it can be hard or woody. Thus if you soak a monkey nut you will soon see that this is really a woody fruit with two seeds inside, and if you take the seeds to pieces you will find them very similar to the peas you have already looked at. An acorn is really a single-seeded fruit and if you break the outer fruit wall off, you will find the seed inside and you may be able to find the small radicle and plumule at one end of it (Fig. 4). The great bulk of the seed is made up of the two cotyledons. A hazel nut is much the same as an acorn, though in it the cotyledons stick together and are harder to make fall apart. There are, however, 'seeds' which are really fruits; thus a packet of beetroot or spinach beet will contain rough rather prickly structures which can be broken up to show small shiny seeds. The one-seeded fruits of the sunflower are like this, and in all the cereals (grasses) there are one-seeded fruits in which the fruit and seed wall have completely fused together. Maize is an excellent example to study, but in many grasses not only have seed and fruit wall joined together but other parts of the flower remain attached to the fruit when it leaves the parent plant, so that altogether it may be quite a complicated structure.

In berries the whole of the fruit wall is fleshy and the seeds are ultimately free in the flesh; the tomato is one of the best examples and, botanically speaking, fruits like squashes, oranges and marrows are all berries. Stone fruits (drupes) are different, for here the inner fruit wall is woody, and the seed is within this stone. Almonds and plums are excellent examples; crack the stones in each case to find the seed. Other fruits split into parts which each contain a seed. The fruits of the parsley family will all split into two halves, while the fruits of the hollyhock divide into many portions each containing one seed (Fig. 5).

Variation in the seeds is also great, though the structures are harder to make out in small seeds. Seed coats themselves are different, for some are hard and shiny while others are slimy and mucilaginous. It is worth while soaking some seeds of linseed and watching the testas swell with the help of a microscope (see Chapter 17). Great swelling occurs in the cells of the testa and the mucilage protrudes in long columns with many spirals of thickening derived from the cell walls among them. Seeds of plants like *Gilia*, *Cobaea* and *Collomia* are also most spectacular.

If the seed is endospermous the embryo may well be very small, as in the seeds of the buttercup and parsley families. If the seed is a large one the embryo is not difficult to see. For example, pull to pieces the seeds of the castor oil or morning glory. To do this, peel off the rather tough outer coat and you will find a thin

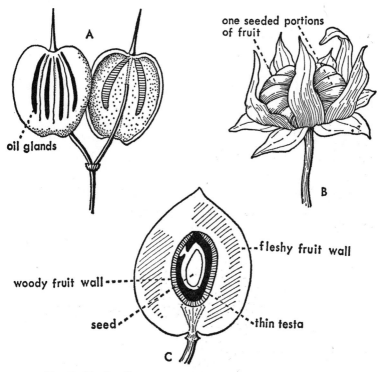

oil glands

one seeded portions of fruit

fleshy fruit wall

woody fruit wall

seed

thin testa

A

B

C

Fig. 5   Fruits of
    A) member of parsley family
        Fruit of a member of the parsley family
        splitting into two one-seeded parts
    B) hollyhock
        Fruit of the hollyhock splitting into many
        one-seeded parts
    C) plum
        Stone fruit of plum

inner coat in the castor oil which is like tissue paper. Peel this off and then gently prise the contents apart along the vertical axis when the two flat, thin, and leafy cotyledons will be exposed to view (Fig. 6). In the morning glory the cotyledons are folded and their shape rather more difficult to make out.

The maize grain shows the arrangement that is found in all the

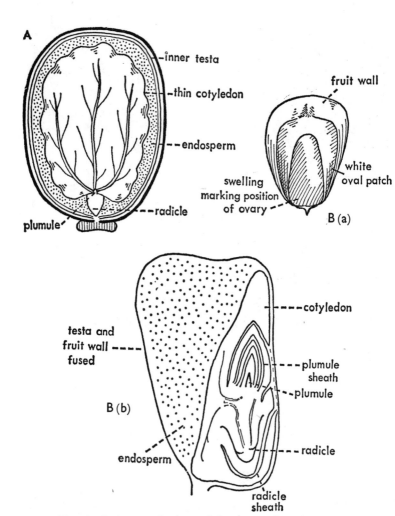

Fig. 6 Structure of (a) *top left*, castor oil seed; (b) *top right*, maize, front view; *below*, maize, left-hand section

grasses. The white oval patch on one side shows the position of the single cotyledon. Look carefully at this and you will see a slight longitudinal swelling running down it. This marks the position of the embryo. Cut the grain longitudinally through the centre of this swelling, and examine one half very carefully with a lens. The embryo is seen to one side of the half, partly protected by a cotyledon; the remainder of the grain is filled with a large food reserve. The tiny shoot or plumule of the embryo is protected by a plumule sheath and the radicle by a similar radicle sheath. Though all this is small, it can be seen readily if the grain has been cut in exactly the right plane (Fig. 6).

Occasionally seeds contain more than one embryo; orange and other citrus fruits have seeds with one large embryo and some smaller ones; these you can easily find by dissecting the pips. Another example is the avocado pear, one seed of which can produce two or more seedlings. After a time one succeeds at the expense of the remainder. The food reserve of seeds is all-important for subsequent germination and growth. Small wonder, then, that it is sought after as food by animals of all kinds. The most frequent reserve is fat or oil, and seeds can be roughly tested for the presence of this by rubbing a cut surface on to paper when it makes a characteristic translucent stain. Other tests for oil include a solution of red dye, Sudan III, in clear methylated spirit. A very thin slice of the seed should be cut with a razor blade, soaked in the dye, when the dye dissolves in any fat present and colours it red. It is most satisfactory to examine the slice under the microscope and actually see the oil globules coloured red, since a vague red colour to the whole section is not conclusive. It is better to crush and boil a number of seeds with water in an endeavour to extract any oil and then test with Sudan III when the fat globules can be seen much better. Alternatively, a few slivers of the seed can be shaken with a clear methylated spirit, which will dissolve the fat. If this is poured off into cold water it forms extremely fine droplets giving a milkiness if fat or oil are present. This is quite a sensitive test.

All seeds contain at least a little protein, and protein turns red if it is heated with a solution known as Millon's reagent. Proteins also turn yellow with strong nitric acid; this colour changes to orange on neutralisation with strong ammonia. (N.B. Care is required when handling these chemicals; also Millon's reagent is expensive.)

The easiest reserve to test for is starch, which turns a dark blue or black with iodine solution. Cut a maize grain longitudinally and just smear the surface with iodine, when the starch reserve will turn black-blue. You can easily compare the starch content of many seeds in this way and make a table to record your results:

Table 2   The food reserves of seeds

Name	Starch	Protein	Fat
Castor oil	Nil	Small amount	Large amount

Notice the distribution of the substance for which you test: whether they are present uniformly throughout the seed and whether they are present in the plumule and radicle.

Occasionally other reserves are found in seeds: sweet corn contains sugar, while the date stores a kind of carbohydrate known as hemicellulose. Of course, too, seeds contain numerous other substances. Some, like the poison coniine of hemlock, are presumably protective in character, while others serve no purpose known to us. They may play some part in the plant's physiology which is yet to be elucidated.

4 Germination studies

THE commencement of growth by seeds is called germination, and it is easy enough to germinate many seeds from packets bought in a shop. Soak them in a little water, keep them in the warm and off they will go. However, it is worth while making a germination test to see exactly how many do grow. One way to do this is sow 100 seeds from each packet you have on damp blotting paper in a

Table 3   Germination tests with seeds

	Date of collection	Number of seeds out of 100 that germinated
Brussels sprout	1969	77
	1968	61
	1965	0
Cauliflower	1969	69
	1968	45
	1967	0
Radish	1969	75
	1968	38
Lettuce	1969	96
	1968	85

saucer, and to count the number that grow within a few days. It is interesting to extend these tests both to old packets of seeds and to the seeds of plants growing in the wild. You will soon find that keeping packets of seeds on a shelf is not a good way to store them, for many lose their power to germinate in these circumstances (Table 3).

When all the seeds germinate at the same time, germination is said to be simultaneous. This happens as a rule in cultivated plants because man has automatically selected varieties with this property by always growing the plants that come up first in any batch sown. If you carry out germination tests with seed collected from wild plants, you will not often obtain simultaneous germination. More than likely you will find intermittent germination over a long period. Sometimes seed germinates better in the second year after sowing, and this you will only find out if you have the

B

necessary patience. In recording germination, daily counts of the number that grow each day should be made, and the results recorded on a diagram (Fig. 7).

There are several reasons why seeds do not germinate immediately on sowing, for many require conditions additional to the

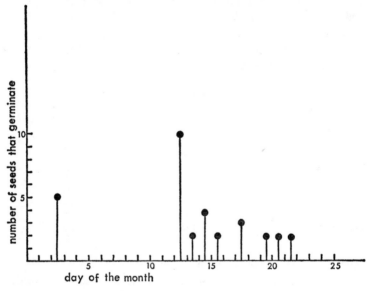

Fig. 7   A record of intermittent germination

three essentials, which are oxygen, water and a suitable temperature. For example, some seeds are light sensitive and will germinate only in its presence while others will only germinate in darkness. Carry out germination tests with as many different kinds of seeds as you have, keeping sets in the dark and in the light. Seeds like rosebay willow herb and curled dock germinate better in the light, but seeds like love-in-a-mist germinate better in the dark. The seeds of *Phacelia tanacetifolia* fail to germinate in the light almost completely. If this kind of experiment is tried under light of various colours it is found that red light is most effective, and it brings about its effects through the agency of a substance in the plant known as phytochrome. Lettuce seed of the variety Grand Rapids will not germinate in complete darkness, but after a small dose of red light moistened seed will germinate. If the red light is followed by far (infra) red light, the initial red treatment is made ineffective. This is called the red far-red reversibility, and this alternation can be repeated a number of times, the response being that of the last treatment given.

Many seeds, especially those of the pea family (Leguminosae), have very thick coats, and uptake of water is not possible until the coat has been cracked, split or rotted in some way. Take seeds of lupin, sweet pea or, best of all, canna, and weigh about twenty of them and soak them in water for twenty-four hours. Re-weigh

Table 4   The uptake of water by seeds

	Percentage increase in weight after soaking for three days	Percentage increase in weight after filing and soaking for three days
Garden pea	250 approx	250 approx
Canna	0	27
Sweet pea	116	233

them, and you will find no difference or only a slight increase. Then repeat the experiment but file a nick in each before soaking the seeds. Rapid water uptake follows (Table 4). The differences in weight in this and the following experiments involving weighing can easily be measured with a small letter balance or similar instrument. You can see in the thick coats a biological device which prevents simultaneous germination, since in nature seed coats rot at different rates according to where they fall. Thus the competition between seedlings is decreased.

Other seeds will not germinate until they have passed through a cold period in the moistened state. Try the effect of soaking some apple pips, hawthorn stones and other seeds, putting them in a refrigerator before germinating them, keeping sets as controls, of course. Again, you can see a biological advantage in this for the plants of temperate climes, for the built-in necessity for exposure to a cold period postpones germination from autumn to spring when conditions are more favourable.

You may wonder why seeds do not germinate inside the pulp of a berry, for apparently moisture, oxygen and a suitable temperature are present. The reason in many cases is the presence of inhibitors, i.e. substances that stop germination. Quite a number of these substances are known. One is coumarin, and you can easily try the effect of this by adding a one per cent solution of it to 100 cress seeds on damp blotting paper in a saucer, keeping a control set to which an equivalent amount of water is added (Table 5).

The pulp of many fruits has such an effect and you can devise many experiments for yourself, trying the effects of the juices of berries and fruits on the germination of cress and other seeds. It

Table 5   The effect of 1 % coumarin on the germination of seeds

	Number out of 100 that germinated	Number out of 100 that germinated after treatment with 1 % coumarin
Cress	95	3
Lupin	40	0
Radish	75	1
Statice	6	0

is believed, too, that not only do fruit juices produce substances that inhibit germination but that some roots can produce substances which stop the growth of seeds nearby. The seeds of some parasitic plants like the broomrape will only germinate in the immediate neighbourhood of their host, and here it has been shown that the roots of the host produce substances which stimulate the growth of the parasite. There is an African relation of the broomrape known as witchweed which is quite a serious pest of cereals. If research could discover and produce the stimulating substance sufficiently cheaply it could be applied to the soil so as to induce premature germination of the parasite's seeds. They would then die, since the host plants would not be available. In this way a serious parasite might be eliminated.

One of the first steps in germination is the intake of water. This can easily be measured by weighing fifty dry peas and soaking them for twenty-four hours and then weighing them again. You can also measure the increase in volume with a measuring cylinder, and if you can do it sufficiently accurately you can try to answer the question as to what extent the water fills up spaces in the seed or is responsible for direct enlargement. If the seeds increase in weight by, say, 50 gm, then the volume increase could be 50 ml if all the water goes to increase it, but if some fills up empty spaces the overall increase will be less.

The uptake of water causes formidable pressures. Try filling a plastic bottle with dry peas and then adding water, or fill up a can to a marked level with dry peas, add water so that they swell, and find the weight required to squeeze them back to their original level. Or you may be able to put together the more elaborate apparatus as shown in the drawing (Fig. 8).

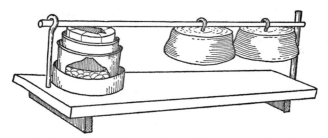

Fig. 8   Apparatus for testing swelling power of seed

How does the water get into the seed? Sometimes it gets through a small hole in the testa called the micropyle. You can find this in a bean just above the scar of attachment. In a runner bean the micropyle is at one end of the hilum and there are two tiny swellings at the other end. The micropyle can be stopped up with one of the many adhesives you now buy in the shops. Soak about twenty beans with stopped-up micropyles in water and compare the water uptake with an equal number of normal beans by weighing them before and after soaking (Table 6). Compare this result with that obtained in Table 4.

Following and coincident with the uptake of water is the mobilisation of the food reserve for use by the growing plumule and radicle. Starch is present in the cells as a tiny grain, and it cannot move about the plant as such; and in fact most food reserves are immobile in the plant. To use them, they must be made soluble, and this involves chemical changes. In living organisms these chemical changes are brought about by enzymes which are akin to the catalysts of ordinary chemistry. Starch, for example, is turned to sugar by the enzyme diastase.

To show the action of diastase, germinate some barley grains for about a week till the radicles are just visible. Take a small handful, cover them with water and pound them thoroughly. Here a pestle and mortar are useful, but it can be done with a large spoon and a basin. Then filter the liquid through a filter paper; the filtrate will contain the enzyme extract. Prepare a dilute solution of starch by weighing out half a gram and mix to a very smooth

Table 6   The effect of stopping up the micropyles of runner beans on the water uptake

	micropyle stopped up	micropyle normal
Weight of beans	8·21 gm	6·37 gm
Weight after soaking twenty-four hours	14·88 gm	17·21 gm
Percentage increase	80%	140%

paste with a little water. Then add 100 ml of hot water, stirring vigorously. Take equal quantities of this solution, and add some of the enzyme extract to one tube and an equal quantity of water to the other. Let the tubes stand for ten to fifteen minutes and add a few drops of iodine solution to each. In the tube containing the enzyme the starch will have disappeared and the iodine colour will remain. The control tube will still contain starch. If possible, carry out a further test with Fehling's solution on the tube containing the enzyme, which will show the presence of sugar. Thus the enzyme turns starch to soluble sugar which can be transported to the growing points of the tiny plant. Similar changes occur in the proteins and oils, and so the materials to build new cells are made available.

Most of the sugar is oxidised in the growing points to provide energy. Some of this is liberated as heat and some used for the actual processes of cell extension and growth. Only a small fraction is used in the making of new tissue. It is possible to show

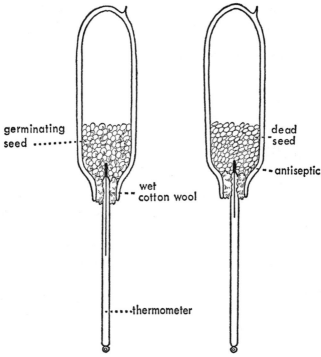

germinating seed ....

wet cotton wool

dead seed

antiseptic

....thermometer

Fig. 9 Experiment to see if heat is given off by germinating seed

that heat is given off by seeds when they germinate if you can obtain two thermos flasks and two reasonably sensitive thermometers. Soak up some peas and pack each of two thermos flasks about one-third full. To one add a small quantity of a disinfectant; this will ensure that the peas are dead but do not decay. Turn both flasks upside down, pack the mouths of the flasks with cotton wool, insert the thermometers so that the bulbs are nicely in the body of the peas (Fig. 9). Energy is liberated by the burning up (oxidation) of the sugar and this is the reason why the seeds require oxygen to germinate. The sugar itself is converted into carbon dioxide and water. The carbon dioxide is given out to the atmosphere (devise an experiment to show this), while the water remains within the seed and contributes to growth extension. If, however, the seed is dried in an oven at 100°C to constant weight, it will have lost in dry weight. Read the thermometer carefully and observe the differences between them.

5 Seedlings

THE cotyledons (or seed leaves) can behave in two ways when a seed germinates; they can remain below the ground all the time or they can come above the ground. The first type is known as hypogeal and the second as epigeal. This difference can occur in closely related plants. Thus the runner bean has the cotyledons below the ground while in the french or kidney bean they come above the ground. We have also seen that the flowering plants can have either one or two cotyledons, and that all seeds may or may not be endospermous. We have therefore the rather complicated set of possibilities:

Monocotyledons
   1. Epigeal   (a) endospermous, e.g. onion
                  (b) non-endospermous, e.g. water plantain
                     (*Alisma*)
   2. Hypogeal (a) endospermous, e.g. maize, date or coconut
                  (b) non-endospermous (very rare) cape pond-
                     weed (*Aponogeton*) only
Dicotyledons
   1. Epigeal   (a) endospermous, e.g. castor oil, morning glory
                  (b) non-endospermous, e.g. cress, marrow
   2. Hypogeal (a) endospermous (very rare), e.g. custard apple
                  (*Annona*)
                  (b) non-endospermous, e.g. runner bean,
                     garden nasturtium

Some of these types of seed can be easily obtained. Of the endospermous monocotyledons with an epigeal type of germination, onion is readily obtained and easy to grow. The little black seed, if soaked, can be cut and readily examined with a hand lens when the embryo will be seen coiled up in a mass of endosperm. The seeds germinate readily enough on damp blotting paper in a saucer when the radicle is the first part seen to emerge. This anchors itself in the ground and shortly afterwards tiny lateral roots are formed. The cotyledon which at first is difficult to distinguish from the radicle now lengthens, but the tip remains embedded in the seed, and so the whole structure is gradually pulled above the surface of the ground. The cotyledon turns green and functions as a normal leaf, absorbing carbon dioxide from the air and thus carrying out photosynthesis. This change in colour helps to distinguish it from the radicle. The cotyledon also functions as an absorptive organ for the tip serves to absorb the food from the endosperm in the seed and pass it on to the growing tissues (Fig. 10).

It is not so easy to find a monocotyedonous seed which is non-endospermous and which has epigeal germination, though it occurs in a group, many of which are water plants. The best known members are the water plantains and the greater water plantain is a fairly common plant in Great Britain, so you can search for it along the sides of streams, ponds and rivers, collect its seed and try to grow it in the following season. The seeds do show a considerable dormancy, but abrading and chipping the seed coat is an aid to germination (Fig. 11).

All the grass family are endospermous and show hypogeal germination. Maize is an excellent example, particularly the larger fruited varieties, and we have already mentioned it in Chapter 3. When it germinates, and it is easy to grow, the first part to come above ground is the hard white pointed plumule sheath. This is a hard tube through which the first leaf grows and unfolds (Fig. 12). This is an interesting example of one of the ways in which the delicate growing point is protected in its passage through the soil. In the runner bean the shoot is bent over as it grows through the soil and it does not straighten until it reaches

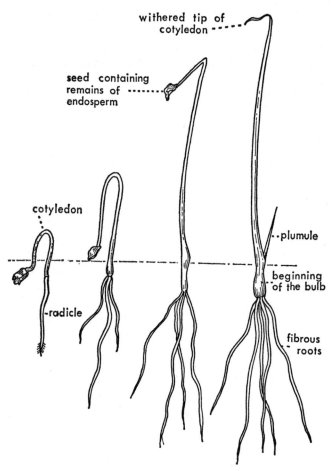

Fig. 10   Germination of the onion

the light. This is another way in which the growing point is protected. During the whole time of germination of the maize the cotyledon remains below the ground and serves to absorb the food reserve from the endosperm and to pass it to the growing root and shoot.

Another large group of flowering plants with a rather similar

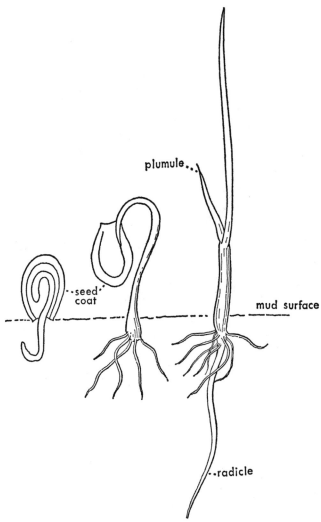

plumule

seed
coat

mud surface

radicle

Fig. 11 Germination of the water plantain

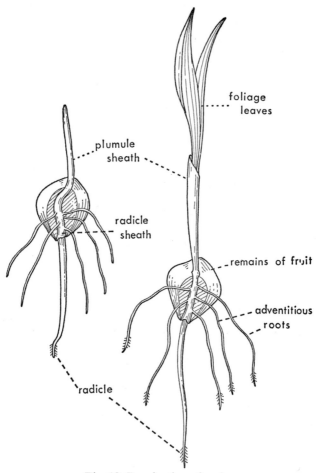

foliage
leaves

plumule
sheath

radicle
sheath

remains of fruit

adventitious
roots

radicle

Fig. 12 Germination of maize

pattern of germination is the palm family. Stones (seeds) of the date can be germinated quite readily in a really hot place, say on a radiator. The mass of the stone is a very tough endosperm consisting largely of a material known as hemicellulose laid down as a thickening to the cell walls. The embryo is small and centrally placed. When germination commences, the tip of the cotyledon enlarges to a large mushroom-like mass which invades, dissolves and absorbs the endosperm and passes it along a stalk or middle piece to the other end of the cotyledon which surrounds the developing shoot (plumule). The extension in length of the middle piece forces the shoot outside the seed where it is in a position to reach the soil surface.

The coconut is basically similar, for as one sees it in the shops it is a one-seeded fruit, the seed being enclosed by the very thick woody shell. One of the three depressions at the end of the shell is thin enough for the seedling to force its way out. The embryo is small, and the white of the coconut together with the milk make up the endosperm. In this seed the cotyledon enlarges to a mass as big as an orange and it absorbs all the nutrients for the growing shoot to use. The food reserve may last for several years so that the absorbing end of the cotyledon remains active for a long time (Fig 13.). The middle piece is quite long so that the actual seedling may be some distance from the nut. It is not too

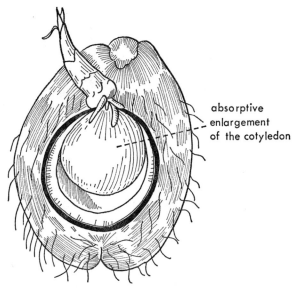

absorptive
enlargement
of the cotyledon

Fig. 13 Germination of the coconut

difficult to germinate a coconut provided the space necessary in a well-heated greenhouse can be found. It may be necessary to try more than one, for, like all other seeds, not every one is viable.

This leaves only one group of monocotyledonous seeds to be mentioned, namely those which are hypogeal and non-endospermous. This combination is a very rare one and it is known only in one genus, *Aponogeton*. There are several species, the most familiar being the cape pondweed (*A. distachyus*), which is sometimes planted in Britain in lakes and ponds (Plate 10). The flowers are white with black anthers and have a scent reminiscent of hawthorn. Seed is relatively difficult to come by, but if you do manage to obtain any, it must be germinated on mud in shallow water. In this seed the cotyledon is a conical structure with the embryo to one side and enclosed in the seed coat. The whole seed floats away from the plant, the seed coat bursts, the contents fall out of it, and sink into the mud where germination takes place. The cotyledon supplies the food material to the shoot at its base which gives rise to the leaves of the new plant (Fig 14).

Turning to the dicotyledons with epigeal germination we are on far more familiar ground and there are many to choose from. The castor oil and morning glory are splendid examples. Both these are readily obtained and fairly easily grown, though the castor oil sometimes germinates badly. In the castor oil the radicle emerges and anchors itself in the ground, and then, by subsequent growth of the part remote from the tip, the whole seed, testa and all, is carried above the soil surface. Since it is quite a large seed, it breaks the surface like a miniature earthquake. The cotyledons function for a time as absorbing organs, passing on the food reserve to the growing point, but later the seed coat drops away and the cotyledons unfold to serve as the two first green leaves. These are more or less oval in shape, whereas the mature leaves are lobed and pointed (Fig. 15).

In the morning glory the endosperm is jelly-like and the embryo large, but the seed germinates in much the same way as the castor oil. Here the cotyledons are lobed, almost butterfly-like. Usually the shape of the cotyledons is simpler than that of the mature leaves and in fact cotyledons do tend to look a bit alike. This may be related to the uniformity of their environment and their similarity in function. However, this is not completely true, for in the genus *Ipomoea*, which includes the morning glory, the cotyledons are lobed in various ways, whereas the adult leaves are heart-shaped. In the buttercup family the cotyledons are often pointed,

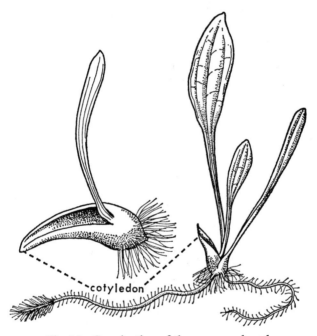

Fig. 14   Germination of the cape pondweed

in the parsleys and carrots they are strap-shaped, despite the finely divided mature leaves, while in eschscholtzia they are forked in a characteristic manner (Fig. 16).

Eschscholtzia is an example of a non-endospermous dicotyledonous with an epigeal germination. Other examples include very familiar plants like mustard, radish and cress, all the food material being packed in the cotyledons. In these seeds the cotyledons tend to remain within the testa until almost all the food is exhausted. Some of these have cotyledons with unusual shapes, like cress with three-lobed cotyledons and clarkia with cotyledons that are broad and short but which later elongate and form basal portions that are more like the normal mature leaves.

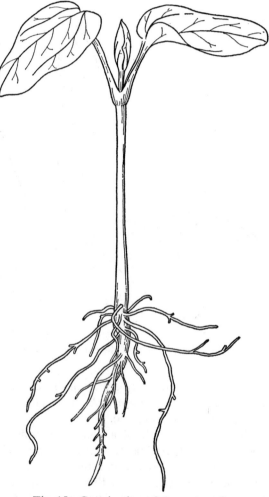

Fig. 15   Germination of the castor oil

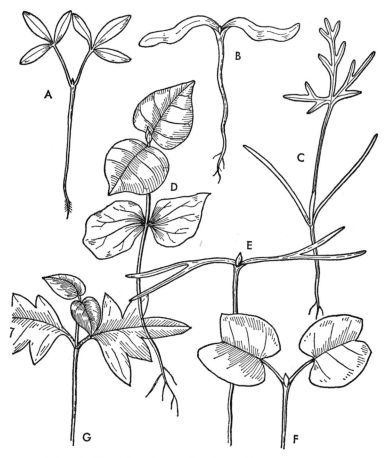

Fig. 16 Cotyledon shapes of some seedlings: A, cress; B, sycamore; C, fennel; D, beech; E, Eschscholtzia californica; F, Ipomea purpurea; G, lime

Other seeds that come into this group are the marrow, squash and cucumber. The seeds are flat and the radicle has a small projection on its inner side which catches the lower edge of the testa. The elongation of the plumule above this forces the upper half of the testa to split and thus aid the emergence of the plumule.

This can easily be seen in any germinating sample of this seed (Fig. 17).

We are left with the group of dicotyledonous non-endospermous seeds which are hypogeal. This is a large group and includes the

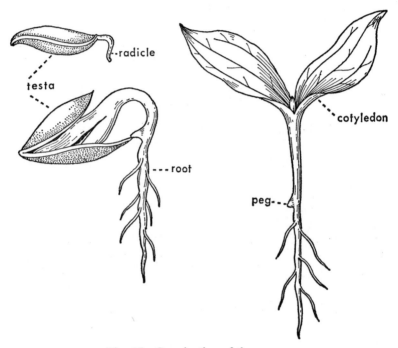

Fig. 17   Germination of the marrow

most familiar runner bean. In it, the food reserve of the cotyledons is transferred down the very short cotyledonary stalks to the growing shoot and root; the plumule grows to the surface with the apex bent over like a shepherd's crook and it unfolds in the light. You can make a box with one side of sloping glass, fill it with soil and sow a row of beans against the face of the glass. In this position the plumules will straighten under the influence of the light but they will find it difficult to force their way through the soil to the surface. In consequence they grow tortuously and some may never reach the surface at all (Fig. 18).

Finally, there is a class of endospermous dicotyledonous seeds

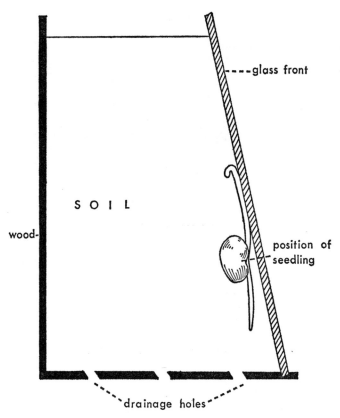

Fig. 18   L.S. of glass-fronted box

with a hypogeal type of germination. This group is very uncommon and the members are all tropical plants. One is the rubber tree and the others include the custard apple, the soursop and the sweetsop, all species of *Annona* (Fig. 19). Curiously, then, endospermous seeds with hypogeal germination are common

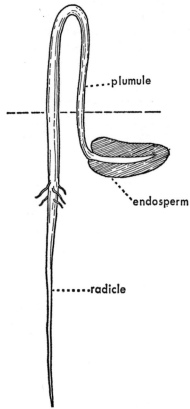

Fig. 19   Germination of Annona

in the monocotyledons, rare in the dicotyledons, while the reverse
is true of non-endospermous seeds with hypogeal germination.

6 Transplant experiments

THERE are many ways of propagating plants, of making two or
more grow where before there was but one. For example, you

can propagate by cuttings, by offsets, by grafts, and you can grow new plants from seeds. There is a fundamental difference between the last and the previous methods in that seeds are the product of a sexual reproductive process, and each seed gives rise to a new individual. In the others a new individual is not produced, but merely a piece of the original is rooted. When you strike a geranium cutting you take a piece of stem and shoot, stick it into the ground, and it develops roots. In a similar way, all the Cox's orange pippin apple trees of the world have been derived from the original tree by grafting, for this variety does not come true from seed. The same is true for hundreds of our garden plants. Such varieties are called clones and the derived pieces are ramets of the clone. The great advantage of this method of propagation from the gardener's point of view is that all the plants are alike. Thus he can plant a flower bed with a single variety and expect all the plants to be similar in shape, and to flower at the same time.

Most of the plants mentioned in the previous chapters come true from seed, for man has selected them specially with this end in view. Nature is not always so selective, and very close observation will often reveal differences between one individual and another. It would not be unexpected to collect and grow some wild foxglove seed and to find here and there a white-flowered plant amongst the resulting population. We have seen, too, in previous experiments, how the growth and stature of a plant can be altered by the manurial treatment it receives. Any individual plant is the product of its heredity and its environment, and both affect the final result.

How do we disentangle the effects of these two factors? In previous experiments we have used carefully selected varieties of seeds and assumed them to be all the same in their inherited make-up. Ramets of a clone are alike in their inheritance and by growing them in different environments we can see the effect of the environment acting alone. An easy way to do this is to take a large daisy or plantain and to divide it up into a number of ramets as equal in size as you can. Plant half in, say, a sandy soil and the other half in a different soil of your own choice. Work out a scheme to compare the growth of the two sets of plants. Count the number of fully expanded leaves, the number of flowering spikes, measure the diameter of the rosettes, the height of the plants, and note the duration of the flowering season. If you do this in anything approaching a comprehensive manner you will find some kinds of plant show great changes in size, etc., in response to experimental treatment while others do not. The former are plastic, the others less so (Plate 3).

It can be interesting, if you have the perseverance, to grow some of the seeds of the daisies or plantains in as uniform conditions as you can, and to see and measure the variation in the resulting individuals. Another experiment to try is to collect some plants which look alike in size and shape, for example daisies or plantains from a pathway, and grow them in as uniform conditions as possible and see how they behave. It may be that they have a different heredity and their similarity is due to a stunting environment. If this is so, some will grow large while others remain small. Again, you may collect plants of annual meadow grass from different places and on growing them you will probably be surprised at how different they become.

You may have noticed that plants growing in rather special circumstances are often different from those in a more normal environment. Plants like the common groundsel are often more fleshy when growing near the seaside than inland. On the tops of mountains plants tend to grow more compact in form, with rather larger flowers on shorter stalks. On shingle beaches some plants are prostrate which grow erect further inland. Try to gather some plants from such localities and grow them side by side in similar conditions. Do the seaside plants lose their succulent appearance? Do the shingle plants grow erect? If they do not come to resemble each other, then the differences must be due to heredity. The dwarf prostrate variety of the broom which grows on the shingle near the sea in Britain keeps this habit when transferred to the garden (Plate 4). Here is an example of the action of the environment in selecting from the total inherited variation of the species, two forms, one suited to very exposed situations and the other to more sheltered places. In fact, evolution in action.

7 A little ecology

In earlier days a botanist was often content to name and list the plants he found in any one particular place. In doing so he soon became aware that plants have very well-defined habitats; some are never found outside woodlands, and others only on the tops of mountains. More than that, many field botanists learnt exactly the place to look for certain plants; thus a woodland plant like the green hellebore is only found in calcareous woodland while a particular kind of buckler fern would only be looked for on mountain screes. The experience of a field botanist over a lifetime amounted to almost a sixth sense in his skill in assessing a habitat as a possibility for this or that species. Put the other way round, it means that many plants are only able to grow in circumscribed situations and that their requirements for survival are critical. There are, of course, cosmopolitan plants like the common reed which are able to grow almost anywhere in wet places and the great majority of plants which lie between the two extremes.

Ecology is the science that attempts to understand why it is that plants grow where they do, how they act on one another and how they act on their environment. It may be that some plants will only grow where there are no frosts in winter, or in soils where the water content never drops to a figure below ten per cent. This kind of relationship can only be found out by investigation and measurement of the factors of the environment which is not always easy to do.

However, as a beginning it is clearly essential to know the plants that occur in any place chosen for investigation and to make a list of them. This is hardly the book in which to treat at length the ways in which plants may be named, but briefly it can be done in about three ways. The picture is often the first thing to turn to, and provided the comparison between the specimen and the picture is carefully made, and that the picture is sufficiently detailed to show the finer points, this is a good way of making an identification. Secondly, and perhaps the best way, is to use a flora. The set books contain keys whereby the species can be 'run down', and these direct the user to the most important differences by which particular species may be distinguished. Thirdly, a comparison between the specimen and a dried pressed specimen is also a good method of identification. A collection of dried and pressed specimens is called a herbarium, and it is a good idea to make such a collection for some of the commoner plants in order to get to know them well. But the collection of rare plants is to be avoided at all costs. Leave the unusual and the infrequent for others to see and enjoy.

Exact identification of some groups of plants like the grasses and sedges is far from easy, and only by practice in the examination of the smaller parts of plants can you acquire the necessary skill. Growing two or more plants side by side, as suggested in Chapter 5, enables you to study them over a longer time and so makes them familiar. You can then see if there are any real and constant differences between them.

Lists of plants found in different places are interesting in themselves. By comparing them you may find a number of plants that are at home in a wide variety of situations while there are others that are very restricted. Curled dock may be equally at home on the shingle of the seaside, on roadside and waste ground and on paths in woodlands, as well as many other places, while an orchid may be found solely on one particular area of chalk downland. In some places you visit you may be able to make a very long list, whereas in others the list may be only a short one. This itself is significant and shows how different plant habitats can be.

There are further ways of studying the relationships of plants to their environment, and the remainder of this chapter is concerned with two or three examples. One interesting thing to do, if you live in suitable country, is to walk up a mountain noting the changes in the vegetation as you climb. You can stop every 100 yards and make a note of the dominant plants. There may be woodland or ploughed land near the base, followed by upland pastures, and, if the mountain is high enough, some truly mountain

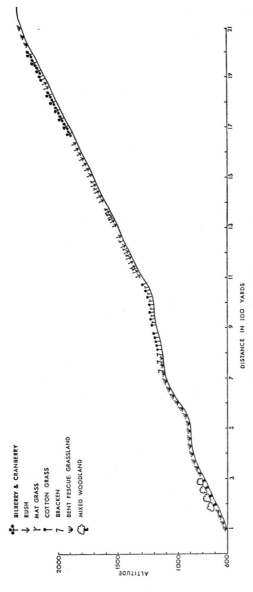

Fig. 20 Mountainside vegetation

vegetation near the summit. Much of this can be shown on a diagram (Fig. 20). The causes of the differences you observe are not so easily defined, for man and animals, climate and soil have all played a part. Climate is obviously different at the top of a mountain. Exposure is so much greater, but at the same time difficult to measure. Wind velocity might be estimated, while humidity and temperature can at least be measured at a given time. Instructive comparisons between the summit and base of a mountain could be made, while the flora of mountain summits could be compared one with another.

Table 7   Light intensity and evaporation in woodlands

Date		open	spruce plantation	Light intensity oak-ash wood	oak-ash coppice
March	13	100	8	71	60
May	2	100	7	25	18
July	7	100	5	12	8
		Evaporating power of the atmosphere (measured with a simple atmometer)			
May	13	100	64	40	
May	20	100	61	40	
July	7	100	34	18	

It may be that you do not live within easy reach of hilly or mountainous country, but woodland may be more easily visited. Here, again, lists will show how different in kind the plants of a woodland are from those of the meadow or road nearby, and a moment's thought will also bring to mind the differences in the environment of woodland and the more open land around it. The interior of a woodland is more sheltered, the extremes of temperature are less and the light intensity is lower. You may realise all this, but not know exactly the measure of these differences. Try therefore to measure some of them, using a simple hygrometer, a thermometer and, for light intensity, a photographic exposure meter (Table 7). You will be surprised at the differences in light intensity, for our eyes are relatively insensitive to changes in intensity above a certain level. Thus the light intensity within a woodland may be as little as one hundredth of full sunlight, and this is highly important for plants since they require

light for photosynthesis. To gain some insight into the significance of light changes, it is desirable to measure the light intensity, firstly as a percentage of full daylight and, secondly to
measure it at intervals over a period as long as a year. You will
then be able to see the drop in light intensity that occurs in a
woodland as the leaves unfold. Just before this event, there occurs
in spring an amelioration of the woodland climate which is related
to the startling growth of the flora at this time. Bluebell, anemone,
celandine and cuckoo-pint are but a few of the plants that con-
tribute to the beauty of woodland in spring. As the leaves of the
trees unfold, the light intensity drops and the number of plants in
full flower diminishes.

Sometimes it is possible to come across the same kind of
plant growing both inside and outside the woodland, and in this
case it is interesting to compare the growth of the two. You may
well find that at a particular date plants like bluebell or bracken
are twice as high in the woodland as outside. When fully grown
there may not be all that difference, but the earlier start has given
the woodland individuals an advantage.

The seaside is a place everyone likes to visit and obviously
it has its own special flora, whether it be salt marsh, sand-dune
or cliff. Sand-dunes are exposed, consisting of fine particles that
are easily blown about, and they are near salt water. Plants that
grow in such places must possess special adaptations to survive.
Marram grass is an important and abundant grass of sand-dunes
and its growth is able to keep pace with the accumulation of sand
around its base. Examine carefully a marram plant and dig down
through the sand of the dune. You will be lucky if you are able to
find the beginning of its root system. Other plants can be explored
in the same way.

Other changes occur as dunes get older. Young (or yellow)
dunes near the sea have a number of plants on them that are
not found on the older (or grey) dunes further from the sea. In
fact, the flora of the dunes changes a great deal from young to
old dunes and you can compare the two by making lists. One of
the factors with which the change is associated is the acidity of
the sand. Sand-dunes near the sea are calcareous owing to the

accumulation of calcium carbonate from the shells of sea animals. As the dunes age, the carbonate is slowly removed and the dune becomes more acid. To show this take similar samples of the sand as you walk inland over the dunes, and add a little dilute hydrochloric acid to each and compare the fizzing of each sample. Try to assess it on, say, a scale of 0 to 5·0 being none at all and 5 the maximum observed. The amount of fizzing is roughly proportional to the amount of calcium carbonate present.

This, of course, is not the only factor responsible for the changes in the vegetation from the young to the old dune, for there are many other changes in the soil, as well as subtle differences in the climate and topography.

8 The work of flowers

To most people the flower is the most important and the most attractive part of the plant; to the plant itself it is the means of sexual reproduction and of the seed dispersal from the resulting fruit. To understand it, choose any fairly large flower and look at its parts carefully. You will usually find a set of green outer parts which protect the more delicate parts while in the bud. These are called the sepals and all together they make up the calyx. Then there are the petals, often brilliantly coloured, which attract the insects for the purpose of pollination, and are collectively known as the corolla. Inside these there are the really essential parts, the stamens which produce the pollen. The pollen gives rise to the male element in the sexual process. In the centre of the flower are the carpels, bearing at their tips the stigmas which receive the pollen. From the carpels a seed box (ovary) is produced which has some means by which the seeds are dispersed (Fig. 21).

There are endless variations on this basic pattern; in wind-pollinated flowers it is not necessary to attract insects, and the flowers are inconspicuous, lacking any large bright petals. In some

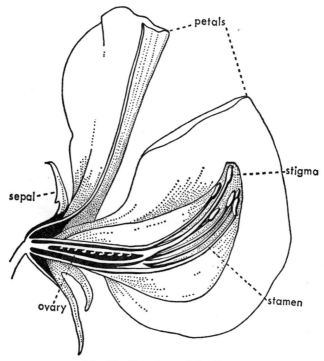

Fig. 21 Structure of the flower

flowers only one set of sexual parts is present; thus flowers may bear stamens only and are said to be male, while in others only the carpels are present and the flowers are said to be female. These kinds of flower may be borne on the same individual or on different individuals. Hazel and oaks have the two kinds of flower on the same tree, but willows, poplars and hollies have the male flowers on one tree and the females on another. This explains, incidentally, why some holly trees never bear any berries. Double flowers seen in gardens are really aberrant ones in which many of the stamens have become converted into petals.

Usually, for seed to be formed, the pollen must be transferred to the stigma. This transference is called pollination and it is brought about by several agencies, the chief being wind or insects.

When the pollen arrives on the stigma it grows a tube which penetrates down to the developing ovule which it enters. Here a sexual fusion between two nuclei takes place and a seed is formed (Fig. 22). Pollen from one kind of flower will normally germinate on a stigma of the same kind of flower, though occasionally it will grow on the stigma of a closely related flower when a hybrid between the two may be produced. Thus the pollen

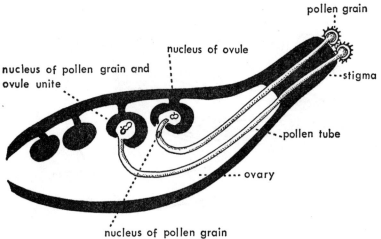

Fig. 22   Growth of the pollen tube and fertilisation

from a common bulbous buttercup will not grow on the stigmas of the meadow buttercup and hybrids do not occur. On the other hand, pollen from the sallow will grow on the stigmas of the goat willow and hybrids between these two kinds (species) of shrub are frequent.

Much of this can be investigated by simple experiment. Make some small plastic or nylon net bags just large enough for the purpose. Then open the buds of some flowers very carefully and remove the stamens with a pair of tweezers, taking great care not to burst them, for this will release the pollen in the flowers and spoil the experiment. Leave some of the flowers to unfold and enclose the others in bags so as to prevent the insects carrying pollen reaching the flowers. These flowers will not set seed, but the others, being accessible to insects, will mostly form seed. For this experiment it is as well to use a plant with a fairly large flower, from which it is possible to remove the stamens without doing great damage.

It is just possible that some of the bagged flowers without stamens will set seed. No pollen has reached the stigmas, so no sexual process has taken place. There are a number of plants which have completely abandoned the sexual process and which are able to form seeds without it. Dandelion is such a plant. With a safety-razor blade slice off the top of the flower head in the bud just above the level of the tiny fruits. In this way the tops of the stamens and stigmas are removed, and though fertilisation is impossible, seed will be set. This operation is a little difficult, because you can cut too low, but a little practice helps. The resulting flower head must be kept moist by being enclosed or covered and should be kept in the shade. You can test the seed for germination very soon after it has formed. This abnormal method of seed formation is called apomixis and it is not so very uncommon in the flowering plants.

The pollen must at least come from the same kind (species) of plant for seed to be set. Plants often show devices to prevent self-pollination, that is pollination of a flower by its own pollen. If a flower bud is tied up in a bag, as previously described, so that insects cannot reach it, it may self-pollinate and set seed. (It may also be apomictic.) Self-pollination can easily be tested for by tying a number of flower buds in bags and comparing the seed set with that in flowers without bags. If possible, count the number of seeds formed in the bagged and unbagged flowers. If there are too many and too small to count they may be weighed if a suficiently sensitive balance is available (Table 8).

There are many devices by which self-pollination and fertilisation are avoided in flowers. In some the stamens ripen so much in advance of the stigmas that they have shrivelled before the stigmas are receptive. The reverse arrangement is also found, whereby the stigmas ripen first, though it may not be so common as the former. One of the most remarkable ways by which self-

Table 8   *The result of removing the stamens
and the bagging of flowers*

Name	Number of flowers from which stamens were removed	Number setting seed
Wallflower	48	0
Sweet pea	8	0
Everlasting pea	10	0
Canterbury bell	24	0
Foxglove	9	0

The result of bagging flowers without removing the stamens

	Number bagged	Number setting seed
Everlasting pea	19	0
Sweet pea	12	12
Canterbury bell	7	0
Garden nasturtium	9	9

pollination is avoided is by what is called self-incompatibility, whereby pollen is unable to grow on its own stigma. In this case a bagged flower may self-pollinate, but no seed is set. This happens in a number of plants, including the garden petunia, and it can usually be demonstrated by bagging some flowers in the bud. Brushing the stigma with pollen from another plant will usually result in seed being set. The mechanism is inherited according to Mendelian rules (Chapter 10), but the results may not always be quite so simple as suggested above.

Most of the experiments and observations described in this chapter are best carried out on insect-pollinated flowers, since these are usually larger and easier to handle than the smaller wind-pollinated flowers. Insects are attracted by the colour and the scent. You can show this, if you have the patience, by removing the petals from some of the flowers in a head or spike and actually watching and recording the number of insect visits to the flowers with petals and those without.

In other cases odour and scent are more important than colour. A very good example of this is the common cuckoo-pint which is pollinated by small midge flies (*Psychoda*) which are attracted by the strong foetid smell emitted by the spadix shortly after the flower spike opens. To show this take a cardboard box, cut a small opening in one side and fit a funnel into this (Fig. 23). Then place

1. Spring sandwort, a plant found in the neighbourhood of old lead workings

2. Hydrangea, a plant whose flower colour is affected by the presence of aluminium

3(a). The greater plantain growing in gravelly soil

3(b). The greater plantain growing in a loamy soil

4. The dwarf form of the broom found near the sea

5. Mendelian segregation for purple and green stem colour in tomato seedlings

6. A bed of dahlias; an example of what the plant breeder can produce

7. Birdsfoot trefoil (*Lotus corniculatus*)

8. White clover (*Trifolium repens*)

9. The Oak gall

10. Cape pondweed (*Aponogeton distachyius*) growing with water lilies

11(a). The action of paraquat: to show that it is inactivated by the soil. Above, one plant is wetted, while paraquat is watered on to the soil round the other plant. Below, three days later, the wetted plant is dead but the plant in the watered soil is still healthy

11(b). The action of paraquat: above, two bean plants are wetted with paraquat. One is covered and the other left exposed to light (centre). Three days later the plant in the light is dead, but the covered plant is still healthy

12. The action of verdone: in each case the left-hand strip has been watered with verdone; above, on dandelions; below, in close-up, on catsear

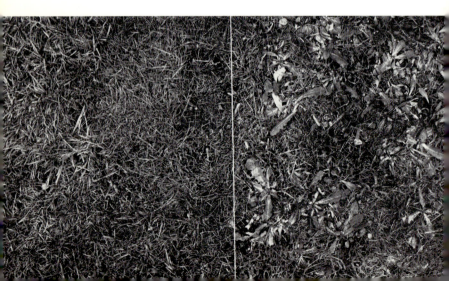

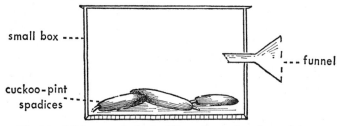

Fig. 23   Midge trap baited with spadices of cuckoo-pint

in the box a number of fresh spadices from cuckoo-pint spikes. The spadices should, incidentally, feel slightly warm to the touch, for at the time that they are most foetid there is a remarkably high respiration rate, resulting in rises of temperature of as much as 10°C. The charged trap should be set in woodland amongst a cuckoo-pint population when it will be found to catch numbers of the midge flies. The trap can be made the basis for further experiments; you can try to use it to find how the number of insects trapped varies with the time of day and how it varies with the kind of weather.

9 Fruit and seed dispersal

SEEDS and fruits are spread about the world in many ways. Some of the most effectively spread plants are almost too familiar, like the burrs that have to be extracted one by one from clothing or the masses of winged fruits of elm and sycamore that block the gutters of the streets. While it is obvious that many devices, like wings, plumes and hooks, do spread plants about the earth's surface, it is not so well known how far they travel or how efficient one mechanism is as compared with another. You can try to answer these problems by painting a hundred of the winged fruits on a sycamore tree (a ladder will be required) and then

C

trying to find them after they have been blown from the tree. This experiment is made the easier if an isolated tree can be found.

An easy way to compare the effectiveness of these mechanisms is to drop the winged fruits of, say, sycamore and ash from the top of a building and noting the time it takes for them to reach the ground. Plainly the longer they remain afloat, the greater the distance they can be carried. You can try not only with the heavier winged fruits like ash, lime and others but also with the very light plumed fruits of thistle and dandelion. Since some of these will remain afloat in winds as light as a one mile per hour breeze you will do better to try the experiment indoors where the air is still.

In wet and damp weather many of the plumed fruits like those of dandelion close up and this means that if they are in the air at the time they fall to the ground at a suitable moment for subsequent development. Try placing some of these tiny parachutes in water to see if they are readily wetted. Try also the effect of covering a nicely expanded dandelion clock with a jam jar that has been moistened on the inside.

Other fruits have hooks of various kinds which adhere to animals' coats. The burdock, cleavers or goosegrass and agrimony are well-known examples. You can make a rough comparison of the efficiency with which these burrs cling by hanging up a blanket and throwing the fruits at them, noting the number that cling to the surface and the number that fall to the ground.

Other fruits are dispersed by being eaten by animals, particularly birds. In this case the fleshy material of the berries is digested by the bird, but the hard seed passes through relatively unchanged. When birds eat grain, a fraction manage to escape being digested. You can investigate this by spreading some droppings of a domestic fowl or wild bird on to some sterile soil and seeing whether any plants come up. Alternatively, you may look at the ground under trees where you have seen birds roosting regularly night after night and see what seedlings are coming up or what plants have grown there. Elder and blackberry are very likely plants to be found.

Some plants have a so-called mechanical means of dispersal, and in these the fruits break open with violence and scatter the seeds. This happens in many of the pea family, in the violets and pansies and in the plants known for this very reason as touch-me-nots. You can easily investigate this by getting some pods of gorse and placing them on the floor of an empty room. As they dry, 'popping' sounds are heard and the seeds are thrown to a distance of several feet. If possible, spread the floor with white paper so you can find the seeds and measure the exact distance to which they have been thrown. Examine also the way the halves

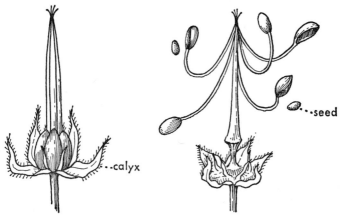

Fig. 24   Dispersal mechanism of cranesbill

of the pod have twisted up and try to work out how this has scattered the seeds. In the cranesbills each of the five seeds is enclosed in a little case attached to a long and elastic strip of tissue. When this swings upwards the seeds are thrown out (Fig. 24). To carry out a fair test of this mechanism, mount the fruits vertically in the centre of the room.

The development of this kind of tissue – tissue that bends or twists according to the moisture content of the atmosphere – can serve another and related purpose. In storksbills, which are closely related to the cranesbills, each of the one-seeded tiny fruits have long spiral awns attached to them which twist and untwist according to the moisture in the atmosphere and in this way tend to bury the seeds in the soil. If you can find any of these

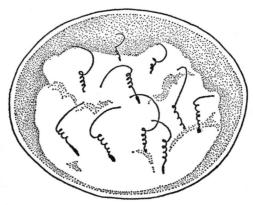

Fig. 25   Storksbill fruits burrowing into cotton wool

fruits (and they are not at all uncommon) place a number of seeds end downwards on damp cotton wool (Fig. 25). You can also try them on the surface of fairly loose soil, which you must water from time to time. Wild oats also do the same thing and so, to some extent, do the garden geraniums.

Some plants growing in water, or by the side of water, have their seeds dispersed by water currents. Usually the seeds or fruits float for a time, after which they sink into the mud below the surface of the water and germinate. You can collect a number of fruits or seeds of such plants and put them in water, and see how long it is before they sink. It may be days or months. Water lilies are good examples to use for this, for the seeds float until the outer part of the seed coat becomes waterlogged. Yellow flag or water iris is another plant to try. Other seeds, like those of the rushes (*Juncus* spp.), usually sink at once or very shortly after putting in water. Later they begin to germinate when they rise to the surface and are dispersed as floating seedlings. There is obviously quite a field for simple experiment here. Try comparing the behaviour of the seeds of the various kinds of waterside plants, and, if you can, compare the behaviour of a waterside plant with the behaviour of a closely related plant which does not grow by the water (for example, two species of dock or two species of umbellifer or two species of willow herb).

One must not forget that man, by his numerous activities, is the most important disperser of all, for by trade and commerce millions of seeds of plants are inadvertently carried about the world. In a modest way any individual carries seeds on his own person. Try brushing out your trousers turn-ups and sowing the dust on damp blotting paper in a saucer. You may be surprised at the number of seedlings that appear. If you don't wear trousers with turn-ups, try brushing the mud from a pair of shoes in the same way. Or go one stage further and try a similar experiment with the dirt from the mudguards of a motor car.

There is yet another way of studying dispersal of seeds and fruits, that is by seeing how many plants can reach special sites, like, for example, the tops of walls or the crowns of pollarded trees. Make a list of the plants growing on the tops of old walls in your district and try to decide by what method they may have got there. Try to discover the nearest individual on the ground from which the seed may have travelled to reach the top of the wall. Or course, quite a lot of seeds may have reached the top of the wall but may have been unable to grow there, so dispersal is not the only factor concerned in determining the flora.

10 Simple genetics and plant breeding

EVERYONE knows that many of the characters of an organism are passed to its offspring; in fact, this is what we mean when we say, for example, that flower colour is inherited. If we buy seeds of the variety of sweet pea known as Gigantic, we expect all the flowers to be white, as indeed they are. But we do know of plants that produce seeds which do not resemble their parents but which throw a range of colours. This is not an accident, but is determined by the laws of inheritance. These laws are well known and in the simplest cases they are easy to show.

These laws were first discovered by an Austrian monk named

Mendel, and he worked with the common or garden pea. This is still a suitable plant to work with and you can buy packets of the following varieties in the shops:

Improved Pilot: This has round seeds, yellow cotyledons, is tall-growing and has pointed pods.

Little Marvel: This has wrinkled seeds, blue or green cotyledons, is dwarf-growing and has stumpy pods.

These varieties are not quite the same as Mendel used but they are more readily available. These varieties will breed true, that is the offspring will resemble their parents very closely, the only differences between them being due to the circumstances in which they are grown. This shows a most important precaution taken by Mendel in his original experiments. Actually the garden pea automatically self-pollinates in the bud before the flower opens, and insect visits are not necessary for seed to be set. Mendel grew his own varieties from seed in this way, and by rejecting any aberrant plants he ultimately obtained stocks which he was certain would be true.

Mendel then proceeded to cross one variety with another and this can be done as follows. It is necessary to stop the process of self pollination and this must be done by removing stamens from the bud while it is quite small. Firstly select a bud, and remove any others near it so that it stands alone. Then, with a pair of pointed tweezers, gently fold back the upper petals that cover the central keel of the flower, and hold them down with a finger. Insert the tweezers so as to cut through one side of the inner petals covering the keel and open up this part of the flower so as to expose the stamens. Examine them closely, and if they have already burst and shed their pollen you must find a bud at an earlier stage and start again. If they have not burst, then remove them with the tweezers, counting them as you go so that you take away all ten, but, of course, you must leave the stigma and style untouched. It is then necessary to return the petals to their original positions and to protect the flower from contamination by foreign pollen. This can be done by covering the emasculated flower with an inverted plastic net or muslin bag and tying the mouth around the stem with cotton or fine string. A little cotton wool may be placed in the mouth of the bag, if necessary, to prevent damage to the plant stem when tying. All unwanted flowers should be removed and also any unwanted pods; this reduces the risk of contamination by the wrong pollen and allows all the resources of the plant to be used to develop hybrid pods. It is a good thing to practise the emasculation on several buds before actually starting the experiment.

After two or three days the stigma will be ready to receive pollen and you can apply some pollen from another variety to the stigma with a tiny paint brush. Tie the flower up again in the bag until the pod begins to form. Then collect the seeds and save until next year for sowing. If you have used the varieties listed above all the seeds will be round and not wrinkled. It is desirable to make several crosses in this way, for so many seeds will be damaged by weevils or taken by birds. In fact, it is a very good idea to net the whole plant against birds.

The collected seeds should be sown carefully the following season a considerable distance apart and the flowers allowed to self-pollinate and set seed. You will find that these plants resemble one of the two parents exactly in respect of most of the characters you investigate. Thus, if you have used the varieties mentioned earlier, all the plants will be tall, like Improved Pilot itself. This Mendel called dominance. Roundness in the seed is dominant to wrinkled and yellow cotyledon to green cotyledon. The colour of the cotyledon is easily determined by chipping a corner of the seed so as to expose the cotyledon beneath the seed coat.

You must collect all the seed you can from these first generation or F1 plants as they are called, and if you sort them out into round and wrinkled, yellow or green cotyledons, you will obtain a ratio of dominant to recessive of three to one. If you sow the seed the following year, spacing the plants so you can tell one from the other easily, you will get a ratio of three to one for tallness against dwarfness. Here are some actual results obtained from a rather small experiment of this kind (Table 9).

It will be noticed that the individual results are often a little way from the three to one ratio but the total for all the five results

Table 9   Inheritance in the common pea

	Numbers	
Tall and dwarf plants	49	21
Tall and dwarf plants	35	16
Yellow and green cotyledons	59	28
Yellow and green cotyledons	99	27
Totals	299	106
Ideal ratio	304	101

is much nearer the ideal. This emphasises that Mendel's results are statistical and only show if large numbers are used. In any one pod all the seeds may be round, but if a large number is counted, the result becomes obvious. Mendel himself worked with large numbers and here are some of his actual results (Table 10).

Table 10   Mendel's results with the common pea

	Numbers		Ratio
Tall and dwarf plants	787	277	2·84:1
Round and wrinkled seed	5474	1850	2·96:1
Yellow and green cotyledons	6022	2001	3·01:1

Furthermore, it is possible to find the heredity ratios for two or more pairs of characters, e.g. what happens when you cross a tall plant with round seeds with a dwarf plant with wrinkled seeds. The results are again statistical and only show up if large numbers are used. The majority of characters are inherited independently of one another and for two pairs of characters in the second generation we get a ratio showing:

9 with both dominants
3 with one dominant
3 with the other dominant
1 with both recessives

Mendel records

315 plants with round seed and yellow cotyledons
101 plants with wrinkled seed and yellow cotyledons
108 plants with round seed and green cotyledons
 32 plants with wrinkled seed and green cotyledons

In two experiments I obtained

> 375 plants with round seed and yellow cotyledons
> 134 plants with wrinkled seed and yellow cotyledons
> 121 plants with round seed and green cotyledons
> 61 plants with wrinkled seed and green cotyledons

The ideal numbers would have been 388, 130, 130, 43

You can also attempt a back cross, i.e. a cross between the hybrid and the parent. Thus the hybrid between Improved Pilot and Little Marvel when crossed with Little Marvel will show approximately equal numbers of round and wrinkled seed.

It is possible to repeat these experiments with other plants, though you will find it best to work with plants having large flowers since these are easier to emasculate. It is also desirable to use rapidly and easily grown annuals so as to get your results quickly. Try studying the inheritance of doubleness in the flower of the garden nasturtium by growing the varieties Tom Thumb (single flower) and Golden Gleam (double flower) and crossing the two. Single is dominant to double.

If you are interested in growing crop plants, P.P.G. of 18 Harsfold Road, Rustington, Sussex, will provide you with packets of tomato seeds together with full instructions for growing and crossing. They will provide you, for example, with a sample of second generation (F2) seeds that will give you a three to one segregation for purple and green stem (Plate 5). This you can see as soon as the seeds germinate, but perhaps not with quite the same satisfaction that you will gain when you do the whole thing yourself from the beginning.

Another nice plant to work with is maize. This large grass requires a long and warm summer to do really well, but it can be grown in England if the grains are sown early in the spring in a warm greenhouse and planted outside in May. Maize is monoecious, bearing the male flowers in a head or tassel and the female flowers lower down in a cob with the stigmas (silks) protruding from the apex at flowering time. Pollination is by wind, large amounts of pollen being produced by the male flowers, which is carried to the silks of the female flowers. Hybridising two varieties then becomes very easy. All you have to do is to grow the two chosen varieties (say one with yellow grains and the other with white grains) side by side and remove all the tassels from the white plants as they form. Thus the only pollen being blown about will be from the yellow, and if it falls on its own stigmas it will give true breeding yellow, but if it pollinates the white the grains will be hybrids. Yellow is dominant to white, so if you take some of the grains from the white plant and sow these side by side the following year cobs with three to one ratio will be obtained. The result may be expressed

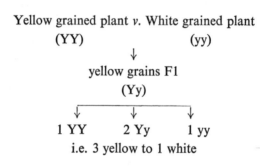

Yellow grained plant *v.* White grained plant
(YY)   (yy)
↓
yellow grains F1
(Yy)
1 YY   2 Yy   1 yy
i.e. 3 yellow to 1 white

If you can get a purple or red grained variety you can grow these alongside the white varieties. Then de-tassel the white varieties as soon as the tassels appear so that no pollen is spread from these by the wind. Allow the tassels of the red grained varieties to flower and you will get cobs on the white plants pollinated by the red and cobs on the red plants pollinated by the red pollen. The results will depend on the exact genetical constitution of the plants concerned but they may be explained by reference to more advanced books on genetics than this.

Perhaps enough has been written to indicate the kind of scope there is for experiment here. Most, if not all, of the varieties of garden and crop plants have been produced by breeding experiments fundamentally similar to those described here. The range of colour in sweet peas, the superb varieties of the garden dahlia, the varieties of the rose are but a few examples (Plate 6). If you do try further experiments for yourself, do not always expect to obtain such simple ratios has have been given here, for genetics is a very complex subject, and there are many complications to the very simple picture that has been presented. You may wonder how many of the inherited variations have come into being. The short answer is that changes in the heredity mechanism occur spontaneously in nature. Man has found means of speeding up the occurrence of these mutations, as they are called; one way is by exposing seeds to radiation. The effect is to cause a number of usually deleterious mutations and you can buy from the Carolina Biological Supply Co, Burlington, U.S.A., packets of irradiated seed. These can be sown and the percentage germination (see Chapter 4) determined, which must be compared with that of normal seed when the deleterious effect of the radiation will be evident.

11 Introductory plant physiology

SOME of the earlier experiments mentioned in this book will have shown that water is the first and most important nutrient for plants. Starve a plant of water and it fails to grow at all, but with a plentiful and steady supply full development of the whole plant is possible. It is true, some plants like cacti grow in deserts and survive long periods of drought, but their growth is very slow compared with the typical plants of our gardens.

Plants pass the water through their bodies and lose nearly all of it to the atmosphere by evaporation through their leaves. The amounts lost in this way may be large. Stephen Hales, one of the earliest plant physiologists, writing in 1727, recorded that a large sunflower plant with a leaf area of 39 square feet gave off about one pint of water in the twelve hours of a hot day while a cabbage with about half this leaf area gave off about one and a quarter pints. Others have considered that a birch tree might give off 400 kilos of water (about 90 gallons) on a sunny day while a really large tree may give off a ton of water per hour.

Some plants exude water from the edges of their leaves in a process known as guttation. These plants have water-secreting pores round the leaf edges, usually at the tips of veins, and in early morning drops of water may be seen at these points. Here again the amounts may be large. The taro or coco (*Colocasia antiquorum*), which is an Indian plant with very large leaves (about a metre and a half long by half a metre wide), has been known to produce 200 ml (about one-third of a pint) from a single leaf in one day.

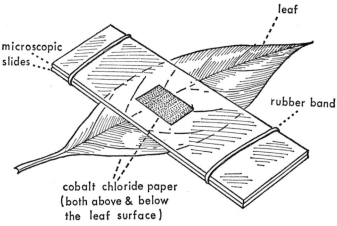

leaf

microscopic slides

rubber band

cobalt chloride paper
(both above & below
the leaf surface)

Fig. 26   The use of cobalt chloride paper

An easy way to show this great water loss is to use paper (filter or white blotting paper) soaked in a three per cent solution of cobalt chloride. When really dry this paper has an intense blue colour but it changes to a pale pink when it becomes moist. If a strip of this is applied to a leaf surface and held in position for a short length of time, it rapidly becomes pink. The more rapid the loss of water (or transpiration, as it is called) the faster the change of colour. If strips of this paper are fastened on to the surfaces of leaves and held in position by glass microscope slides (or something similar) a rough comparison of the rates of transpiration can be made by noting the time it takes for the colour to change. It is essential that the paper be dry in the first place and that it is rapidly placed in position, for the moisture in the atmosphere soon induces changes (Fig. 26).

More elaborate methods exist for measuring this water loss. One is to take a whole plant, roots and all, and place it in a bottle of water. The surface of the water is covered with a little oil to prevent evaporation from this water surface and the level carefully noted from day to day. As the water is absorbed, so the level falls, and the actual amount may be measured by refilling the bottle up to the original level from a measuring cylinder. This experiment really measures the rate of water uptake by the roots, which is nearly the same as the amount lost by the leaves.

The actual path the water follows from the root to the leaf may be seen if a garden weed such as groundsel is dug up, its roots washed free of soil and placed in water coloured with red ink. In an hour or so the red ink will pass into the plant with the water, colouring the tissue through which it passes. If the root, stem and leaf are cut across at various places and examined with a lens, the conducting tissue, called xylem, will be seen. This tissue is placed mainly centrally in the root, but towards the outer edge of the stem and in the veins of the leaf. Seen under the microscope the xylem consists of fine long tubes with few cross walls, but with strongly thickened side walls. (Fig. 27)

As seen in previous chapters, plants gain not only water but mineral salts from the soil. From the air leaves gain carbon

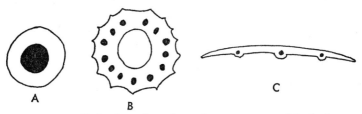

Fig. 27   Stained sections through root, stem and leaf of groundsel plant immersed in red ink (shaded portions indicate path of transport)

dioxide, and with the aid of light they build carbohydrates, the most important of which is, and, incidentally, the most easily detected, starch. Starch gives a blackish-blue colouration with iodine, and it is quite easy to test some parts of plants like tubers, rootstocks and seeds for starch by cutting them open and smearing the freshly cut surface with iodine. You can work through a number of vegetables like potatoes, artichokes, parsnips or fruits like bananas or seeds like beans in this way. Nearly all contain starch but some plants make reserve compounds of a slightly different character which do not turn blue with iodine.

The starch is first made in the leaves, for light absorbed by the green pigment, chlorophyll, supplies the energy for this process. It is not possible to test for the presence of starch in a leaf by the addition of iodine for the simple reason that the presence of the chlorophyll masks the blue colour of the starch-iodine reaction. To test leaves for the presence of starch it is necessary, first of all, to remove the chlorophyll. This may be done by boiling the leaf in methylated spirit (or isopropyl alcohol). Immerse the leaf in the alcohol to which a little water has been added, in a boiling tube, and then warm the tube in a saucepan of hot water, taking great care that the alcohol does not catch fire.

A few minutes' boiling in this way will usually remove enough chlorophyll for the starch test to be used but if the leaf is thick decolouration may take longer. It is better, in the first case, to work with thin and fast-growing leaves which are better suited to this kind of experiment. After boiling in alcohol, the leaf will be stiff and hard. In order for the iodine to penetrate easily, soak the leaf in warm water, when it will soften and turn limp, and then add the oidine. Test a number of leaves in this way and make a list of those which form starch and those which do not. You will find that variegated leaves which lack chlorophyll in some parts do not form starch in those areas. Many brightly coloured leaves form starch, but in these the presence of chlorophyll is masked by additional pigments.

Various kinds of sugars occur in plants; in fact, they are readily turned into one another and easily combined to form starch and other reserves in the plant cell. The most important sugar is glucose and this has the property, also possessed by some other sugars, of bringing about a chemical change known as reduction, in certain chemicals, particularly Fehling's solution. Since this reaction is important in medicine, it has become a standardised test and the necessary tablets can be purchased under the name 'Clinitest' tablets.

Take half a small apple and cut it into small pieces and boil it in water sufficient to cover it. Add one tablet and observe the result, which is a positive test for reducing sugar. This test can be tried on numerous plant extracts and the sugar content compared in this way. Thus plants take in carbon dioxide from the air and with the aid of light make a range of compounds like sugars and starches. These serve for food for animals of all kinds and a moment's thought will show how important the process is, for without it there would be no animals and no mankind.

The making of starches and sugars is but the beginning of the plant's synthetic activity, for from them the plant goes on to make a vast range of products which puts man's chemical factories completely in the shade. Thousands of different chemical compounds are made, complex proteins of which life itself is constructed, pigments to colour petals, celluloses to strengthen cell walls and so forth. Many of these are of great use to man for they provide useful medicines, flavourings, oils, mucilages, beverages, and so on, while the larger tissues provide us with timber and all its associated products.

Everybody knows that plants respond to various stimuli. We say 'they grow towards the light', but really it is only the tips of the main shoots that respond in this way. Roots react differently, being relatively or completely insensitive to light, yet responding to gravity. One of the most exciting responses is shown by plants that are sensitive to touch. The sensitive plant (*Mimosa pudica*) is a very striking example and seed of it is fairly easy to obtain. It germinates well in a warm place (preferably a greenhouse) and can be grown quite easily. When touched the leaflets fold up, and ultimately the whole leaf droops. After a lapse of time it recovers and the leaves unfold again.

Take a well-grown and well-watered plant, and pinch one of

the top leaflets with a pair of tweezers, and notice, firstly, how the leaflet folds up and, secondly, how the stimulus is passed to nearby leaflets. Allow the plant to recover and stimulate it again in the same way and note the time as before. Try to see if there is any evidence of 'fatigue' in the response. There are other experiments you can try. See, for example, how the response is affected if the plant is placed in a much cooler situation and try the effect of covering the plants so they are in the dark. It is, by the way, important to avoid jarring the plants when moving them.

Even more exciting is the movement of the Venus's fly-trap (*Dionaea muscipula*), a plant confined to swamps in the southern United States. It has to be grown in peat, but plants can be bought from nurserymen and with care one will grow in a warm greenhouse. It must not be watered with a hard water since it is calcifuge. The leaf blade of this plant is in two halves hinged at the base so that they can close together. The edges of the leaf blade are set with stiff spines which interlock rather like the teeth of a trap. The inner surface of each part of the leaf has about three long stiff hairs which are exceedingly sensitive for when one is touched the two halves of the leaf close suddenly. The time taken is about ten to twenty seconds, which is fast enough to trap any insect wandering over the leaf.

One very significant point about these sensitive leaves is that they do not react to water drops falling on the sensitive hairs and thus they are spared the consequence of frequent reaction to rainfall.

12 Plant biochemistry

IF you have a small chemistry set, it is quite possible to do a few experiments in biochemistry. In Chapter 11 you have already prepared an extract of chlorophyll in spirit. Chlorophyll is one of the most wonderful substances of the plant world for it is the means by which light energy is used in the making of carbohydrates in the plant. Actually chlorophyll, as found in most plants, is a mixture of at least four complex substances. Two of these are the chlorophylls a and b. Chlorophyll a is a blue-black solid which gives a green-blue solution in alcohol, while chlorophyll b is a green-black solid which gives a green solution in alcohol. Two other substances, or rather groups of substances, are present, namely the carotenes which are orange-red, and the xanthophylls which are yellow. It is fairly easy to obtain a rough separation of the four groups of pigments. The simplest way is to dip a piece of

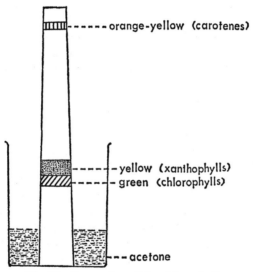

- - - - - orange-yellow (carotenes)

- - - - - yellow (xanthophylls)
- - - - - green (chlorophylls)

- - acetone

Fig. 28   Separation of the chlorophylls

white blackboard chalk into the leaf extract and then allow the green coloured base to dry. Then dip it in the extract again and allow to dry once more. In this way you can build up a concentration in the base of the chalk. When this has been done stand the chalk in a little fresh spirit or acetone (Fig. 28). As the liquid travels up the chalk, coloured bands begin to appear. The upper band is orange-yellow and contains the carotene, below it is a yellow band of xanthophylls, while the two chlorophylls occur at the base. If you have chromatographic materials a better and more effective separation is possible; usually the details of how to do this are included with the materials.

Most plants are green, but, above all, flowers are associated in the minds of all of us with bright colours. Colour in flowers is largely caused by compounds called anthoxanthins and anthocyanins. Anthoxanthins are common in leaves as well as in flowers, and they show a limited colour range. These pigments turn a bright yellow with the vapour of ammonia. Put a few drops of ammonia in a flask and through the neck pass a flower like a snowdrop, white lily or lilac or almost any other white flower. It will turn yellow, but if acid is added the colour will go. With a little ferric chloride solution, a green or brown colour is produced.

To see if anthoxanthins are present in leaves, make a hot-water extract of any leaves you please and carry out the tests already mentioned above. Common plants like docks, dandelion, violet and plantain all contain anthoxanthins.

Practically all the blue, red and purple colours of flowers, fruits and stems are due to anthocyanins. These pigments are found dissolved in the cell sap and it is an exception to find a plant without them. These are readily extracted with some other substances by alcohol; it is best to use flower petals since they have fewer additional substances than leaves. So grind some petals with an eighty per cent solution of alcohol and filter. To the filtrate add a little acid and notice the bright red colour. Add a little alkali to a little of the extract when it will turn green. This shows a property of many of the anthocyanins, namely that they act as indicators, being one colour with acid and another with alkalis. You may remember that litmus, widely used in schools, is such an indicator and that it is derived from a lichen.

It follows that if the acidity of the cell sap changes so will the colour of the dissolved anthocyanin change. This is one of the ways in which flower colour is altered, as, for example, some kinds of morning glory, where the colours change as the flower opens and fades.

Test the alcohol extract also with ferric chloride; if the extract is reasonably pure it will give a slaty blue to a purple colour, but if it contains some anthoxanthin, as is quite likely, it will give a more olive-brown colour. Test the extract also with caustic soda and also with sodium carbonate, and compile a table showing the results. The exact colours obtained will depend on the different anthocyanins and xanthins present in your extracts which again depends on the flowers you have used. Plants which often give results quite different from the usual include members of the Chenopodiaceae (e.g. beetroot), the Phytolaccaceae (e.g. poke-weed) and the Portulaceae (e.g. purslane). The pigments of these are largely soluble in water, in contrast to those first mentioned which have to be extracted with alcohol.

Furthermore, there exist substances known as leuco-antho-cyanins and these produce anthocyanins on heating with dilute (ten per cent) hydrochloric acid. Take a few square centimetres of leaf tissue, chop them very finely and place them in a test tube. Then cover the leaf tissue with hydrochloric acid, and keep the tube warm for about fifteen to twenty minutes. This can be done by placing the tube in a saucepan of boiling water. After this an equal volume of amyl alcohol is added to each tube, the contents shaken and allowed to settle. A red colour in the amyl alcohol indicates the presence of anthocyanidin but a green colour, or hardly any colour at all, indicates their absence. A number of different kinds of leaves may be tested in this way; you will find that plants of the same family tend to give similar results, showing that, on the whole, chemical investigations tend to confirm relationships based on structure.

Another very simple and interesting piece of biochemical work concerns the distribution of what are called cyanophoric gluco-sides in plants. Plants containing these substances release minute quantities of prussic acid when crushed, for example some members of the Leguminosae and Rosaceae like common white clover (*Trifolium repens*) and bird's-foot trefoil (*Lotus corniculatus*) (Plates 7 and 8). The other common clovers do not produce these glucosides nor does the bird's-foot trefoil of marshy places (*Lotus uliginosus*). To test for these glucosides, sodium pic-rate paper is required. This is made by soaking filter paper in a one per cent solution of picric acid, and then allowing it to dry. Imme-diately before use, a strip of this paper is moistened with ten per cent sodium carbonate solution and suspended in the moist condition above the plant tissue which is to be examined.

An easy way to test a plant with this paper is to take a leaf of the white clover, place it in a small tube (say 5 cm by 1 cm), add a drop of chloroform and cork, slipping a piece of picrate paper in between the cork and the side of the tube. In the presence of the glucoside, the paper will turn orange and finally brick red. Results are more speedy if the tube is warmed, but, if this is not possible, the tubes may be left overnight. If chloroform is not available the leaf can be chopped or pounded or otherwise crushed before use.

Using this technique, test several white clover and bird's-foot trefoil samples from different populations of these plants. Both cyanophoric and acyanophoric plants may be found in the same population, and the proportions of the two may vary from place to place. In Great Britain the acyanophoric plants are the more numerous, while the reverse is the case in the south and west of Europe. The acyanophoric plant is the more frost resistant, while the cyanophoric plants are not so readily eaten by rabbits and slugs. In the north and east of Europe frost resistance seems more important to the white clover, while in the south and west protection from grazing confers greater benefit on the species.

The common bird's-foot trefoil is very highly cyanophoric in Great Britain and it has been noticed that where white clover and bird's-foot trefoil grow together there is a lower proportion of white plants with cyanophoric glucoside. Carry out tests to see if this is so in your district. It may be that the bird's-foot trefoil keeps grazing creatures away and thus allows the non-cyanophoric form of the white clover to flourish.

If you are sufficiently patient very interesting experiments can be done to show the difference in palatability of bird's-foot trefoil plants. It is necessary to grow both kinds in a box and then to introduce some slugs. Slugs will feed much more vigorously on the acyanophoric plants. This is a very good example of natural selection in action.

White clover is easily recognised by its white flowers; the two other common clovers are purple and alsike clover (*Trifolium hybridum*). Alsike clover has pink flowers with much longer teeth to the calyx than in white clover. Without the flowers the distinctions are not so easy, but white clover has smooth leaves with very fine teeth on the margins and a creeping habit of growth. Alsike clover does not creep but it does have leaves with very fine teeth. The purple clovers have hairy leaves.

Bird's-foot trefoil has yellow flowers (about five in a group) and what appears to be leaves with five leaflets. (Actually it is trifoliate with two basal stipules which resemble leaflets.) It is common in pastures and grassy places. Marsh bird's-foot trefoil is a similar but larger plant usually found in much damper situations. It produces runners more freely than the ordinary bird's-foot trefoil, and the stem is hollow, whereas in the ordinary bird's-foot trefoil it is solid or nearly so.

13 Plant products

As we have seen, the plant may be compared to a chemical factory producing all sorts of compounds of unending complexity. Just think for a moment of all the compounds we get from plants like foodstuffs of many kinds, flavouring matters of great variety, drugs and medicines, fibres of all sorts, essential oils and scents and so on.

Most important to man are the proteins, compounds that contain nitrogen and other elements linked with the carbon, hydrogen and oxygen of the carbohydrates. These include the most complicated substances known to man and are the compounds, above all, with which life is associated. The proteins occur as reserves in seeds and fruits and it is in this state that they can be recognised by simple tests. Actually no single test works for all proteins, for they are built of amino-acids and the tests used are really tests for one or other of the amino-acids, and any given protein may not contain the acid for which the test you are using applies. However, many proteins give a yellow colour with nitric acid which subsequently turns orange on the addition of ammonia, though if you try this test you must be very careful, for nitric acid is very corrosive. Most of the other tests involve the use of rather more expensive chemicals.

It is often said that bread is the staff of life, so important is it to the diet of the Western world. Bread is made from wheat flour and if you test either the flour or the grain with iodine you will find it to contain a very large quantity of starch. To make bread, flour is mixed with the fungus known as yeast and left to 'work' for a time. The yeast produces carbon dioxide which inflates the dough and when it has finished rising the bread is baked.

This can be shown by an interesting experiment. Take about 5 gm of yeast and make this into a stiff dough with 25 gm of flour (not self-raising). Roll the dough into a short plug on a flat surface and then drop it into an empty measuring cylinder. Any parallel-sided glass or plastic jar will do, but a narrow one is best. If you use a jar, stick a piece of paper down one side, and mark it at equal intervals so as to make a scale (Fig. 29). Immediately on placing the dough into the cylinder or jar, read its height, and do this every five minutes for about an hour, or until it does not rise any more. Then press it down again with a spoon or piece of wood, and see what happens in the next half-hour.

It is a curious fact that if you try to make a dough with flour from other grains, say for example rice, it will not rise. It is remarkable that you can only make raised bread with rye or wheat, but rye bread is darker in appearance and is considered less palatable by most of us. Starch is the common ingredient of all these grains, but individually they differ in their protein content so it is to this that we look for an explanation. Only rye and wheat have the right kind of protein for bread making. The proteins of wheat make a sort of elastic mass with water which stretches as the carbon dioxide produced by the yeast bubbles through it. If you make a dough precisely as indicated above and hold it under a running tap, the starch is gradually washed away as a milky liquid and ultimately you are left with a sticky mass called gluten. To wash away all the flour takes a long time, but

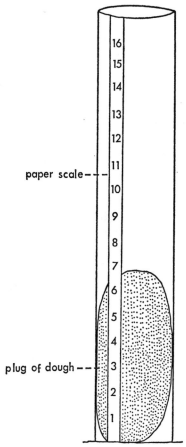

paper scale

plug of dough

Fig. 29   The rise of dough

in fifteen minutes you will notice quite a difference in the dough.
It is the presence of gluten which enables bread to be made.
Moreover, its amount greatly affects the quality of the bread.
Wheat which contains twelve to fourteen per cent of gluten is
known as hard wheat and a relatively dry climate like that of
N. America is needed to grow it.

Wheats grown in milder climates like that of the British Isles contain less protein and are known as soft wheats. They do not make such good bread and are mainly used for making biscuits.

By contrast you can try experimenting with a very minor plant product and make some writing ink. The earliest inks known to Egyptian and Chinese civilisations about 2500 B.C. were made with lamp black and a solution of a glue or gum. This was usually diluted with water before use. Coloured juices of plants were also used but the blue-black ink in use for many centuries was made from plant tannins and an iron salt. You can make your own writing ink in the following way. Find some of the brown spherical galls which attack the oak tree; they have various names, 'oak marble', 'oak nut', 'marble gall', and they are usually more frequent on scrub oak and young trees in coppice than on older mature trees (Plate 9). Pound them finely in a mortar and boil up the powder with a small amount of water, and allow it to stand for a while. Then filter, and the extract will contain tannin. If a little of this is diluted with water and a drop or two of five per cent ferric chloride is added, a deep blue-black ink is produced. If used for writing on paper it may be faint to begin with but on standing it becomes darker and insoluble in water. This insolubility gives the ink its permanence, so that it does not wash out when the paper is wetted.

Many soluble dyes are now used for all kinds of coloured inks. These fade in strong light and wash out in water, and are very suitable when a lasting permanent record is not required. All the same, they will last quite a long time if well protected. Ball-point inks also consist of dyes or pigments dispersed in oils or resins and are really more like printing inks.

If you can't find any oak galls, you can use the bark stripped from two- to three-year-old twigs of the oak, but it should be cut into small pieces for extraction. You can also use the bark of the sumach (*Rhus coriaria*) or the bark of the sweet chestnut. Even the leaves of the latter contain some tannins and can be used. If you use the bark of the horse chestnut and treat the extract with ferric chloride a green colour is given. Walnut and larch also give this reaction but it is necessary to boil well to get sufficient of the tannin extracted.

One other interesting point about the horse chestnut may be studied. It contains aesculin, a substance which belongs to a group of compounds known as saponins, and it gives a remarkably blue fluorescent solution. To see this strip the bark from some young twigs and boil them in a little water. Filter and pour the filtrate into a larger amount of water when the blue fluorescence will be seen.

14 Weed control

THE control of weeds in crops has undergone dramatic changes in the past twenty or thirty years. Cornfields scarlet with poppies, blue with cornflower or yellow with charlock have almost completely disappeared, and many spring or summer annuals of cultivated fields have become rare. All this has been brought about by the use of synthetic herbicides or selective weedkillers. Previously it was very difficult to weed a cereal crop, so rotations were practised and when the land was carrying a root crop it was weeded by hand or by a cultivator. Now all this has changed and cereal crops are sprayed soon after their germination so that many weeds are killed in the seedling stage. Also, by using appropriate sprays, lawns and playing fields can be made free of daisies, buttercups and plantains without the tiresome labour of weeding by hand.

These new methods have been brought about as the result of investigations into the growth of plants by people trying to find out such things as how it is that the main shoots of plants curve towards the light or main roots grow towards gravity. Very briefly it has been found that such curvatures are brought about by the secretion of growth substances by the tips which diffuse backwards and change the growth rates of cells so that one side grows faster than the other producing a curvature. Growth substances are active at very great dilution and it has been found that if one of these substances is used at too great a strength it so distorts the growth that the plant dies. Further, if a growth substance is sprayed on to plants, rather more is absorbed by the flat broad-leaved dicotyledons than by the narrow-leaved upright growing monocotyledons, e.g. some grasses. So it becomes possible to check the growth of dicotyledonous weeds in cereal crops. In the very earliest days of this work the chief growth substance produced by plants was used, but now scientific research has produced a wide range of synthetic weedkillers to suit almost all needs.

To show the effect of one of these selective weedkillers, mark out two similar areas (say a square metre) on a lawn and water one of them with a weedkiller such as Verdone, following the instructions given. You can then compare the treated and un-treated plots at intervals of three days or so, when you will see the effects of the chemical in distorting the growth and causing death. Record the time taken to kill various weeds and note these that are immune or not very susceptible to the weedkiller (Plate 12).

You can also show the effect of the selective weedkiller more precisely if you prepare a normal solution of the Verdone and make successive dilutions of it. You can dilute it five times, then ten times and so on. Then put some of each dilution in a saucer, and in each place ten to fifteen bean or other seeds. Keep the saucers in the dark for a week, and measure the lengths of each of the young roots at three-day intervals. In this way you will find out the dilution which prevents growth and the precise effect on the rate of elongation of the roots of any seeds you choose to investigate.

One of the newest and most striking chemical weedkillers is the substance paraquat which is marketed under the name Weedol. Like most chemicals it should be handled carefully and not allow-ed to splash the skin, and you should be careful to wash your hands after using it. The contents of a tube should be made up according to the instructions, when it can be tried out in your garden. It kills all green vegetation, but it does not affect the bark of trees or shrubs. This makes it specially useful as a weedkiller in rose beds, because if it is skilfully applied round the bases of the trees it will kill the weeds growing close to the roots.

The mode of action of this weedkiller is very interesting. Take two potted plants and spray both with Weedol (Plate 11), but keep one in the dark. The plant in the light will rapidly turn brown and die, but the one in the dark will remain green until you bring it into the light, when it, too, will die. Again, if you grow a plant for some time in the dark and then treat it with Weedol it will remain unaffected even in the light until it develops the green pigment chlorophyll. It is therefore quite clear that there is a connection between the action of paraquat of Weedol and photosynthesis (see Chapter 11) and it is possible to find out more about this by further experiment. Float a broad-bean leaf on a solution of Weedol in the light and the leaf will turn black. You should keep another leaf on water at the same time as a control. Also, float a similar leaf on Weedol solution but keep it in the dark. This last leaf will not turn black. Again, if you float a leaf on a dilute solution of hydrogen peroxide it will turn black, and from this and other more complicated experiments it has been shown that the paraquat of Weedol diverts or alters the process of photosynthesis so as to cause the leaf to make hydrogen peroxide, which then kills the plant tissues.

Weedol is rapidly and strongly absorbed by the surfaces of the leaves of plants, so much so that if it rains soon after the weedkiller has been applied it does not affect the treatment. You can easily test this for yourself by treating plants with Weedol and then watering half of them after an interval of ten minutes and seeing what difference it makes. Equally, the paraquat of Weedol is held very tightly by the soil, and in the soil it is unavailable to the plant. Try watering the soil around a plant with the solution and you will find it has no effect at all. Only on the green leaves does it kill.

To do justice to the substance it does not accumulate in the soil because it is slowly destroyed by micro-organisms. It is unavailable to soil insects or earthworms, so that any cumulative effects from its use are very unlikely.

15 Grafting and budding

THERE are many ways of propagating plants apart from seed; striking cuttings is one, and almost everybody must be familiar with the practice of raising geraniums or chrysanthemums in this way. It has some special advantages for the horticulturalist; for for one thing one can be sure that all the plants will be the same, and so a gardener can produce the 'carpet bedding' commonly seen in parks and large gardens. Taking cuttings is also particularly valuable for those varieties of plants which do not come true from seed.

Plants differ greatly in their capacity to root from cuttings. Most willows root with great facility, for almost any shoot pressed into damp ground will grow. It is also quite easy to root a number of shrubs like fuchsias and roses but it is very difficult to do the same with apples and pears although they belong to the same family as the rose. Apples, pears and particularly roses are known in countless varieties, most of which do not come true from seed. Also, to raise apple trees from seed requires about ten years or more before they will bear fruit and this is one of the reasons why apples and pears are propagated by grafting. In grafting, a shoot known as a scion is shaped and fitted into an incision prepared in the stem of a rooted plant called the stock. A good contact between the growing tissue of the stock and scion is essential for success and the two are bound together and covered with wax so that the tissues do not dry up before they have grown together. By the process of grafting we obtain uniform individuals just as in the practice of taking cuttings, and further, an apple tree becomes capable of bearing fruit in five years or about half the time it would normally take. All the Cox's orange pippins in the world have been derived in this way from the original tree which raised from a pip of the variety Ribston pippin by Mr Cox about 1840.

There is one difficulty as far as apples are concerned, and that is the production of a sufficient number of stocks since cuttings do not root easily. This can be overcome by layering shoots, i.e. covering the basal parts with soil while they are still connected to the parent, but it still takes two years for them to root.

Although grafting assures that the variety comes true to form, it has been found that the kind of stock used will greatly affect the growth of the resulting tree. This was studied by the research station at East Malling in Kent, and so the apple stocks available in England are called after this place. Thus a scion grafted on East Malling stock IV will produce a small and early cropping tree while a similar scion will make twice as much growth when grafted on East Malling XII.

It is a pleasant practice to raise your own apple trees. Dwarfing stocks such as East Malling IV can be bought from the larger nurseries that specialise in fruit trees and they should be planted out and allowed to establish themselves. Then in the autumn following choose a suitable scion which has the same diameter as the stock and cut as shown in the diagram (Fig. 30). The fit between stock and scion should be as close as possible and the two should be tied together tightly with raffia and the union coated with grafting wax.

As a rule you can only graft varieties of one species on one another, e.g. apples on apples, roses on roses, etc. Sometimes you can graft a species on another very closely related one, but that is usually as far as you can go. You can graft apple on pear but it will only grow for one or two years during which time it weakens and dies. Lilac is often grafted on privet, but it does not last very well. Sometimes grafting is used to start the plant off as in some delicate varieties of clematis. These are grafted on to small plants of more vigorous species and planted with the graft below soil level. Then the scion produces its own roots and ultimately becomes independent of the original stock.

Budding is a similar practice; it consists in making a T-shaped insertion in the bark of the tree or shrub and gently opening the cut so that a small segment of the scion bearing a bud can be inserted underneath. This piece must be suitably shaped and must consist only of a piece of bark with a bud so that contact and

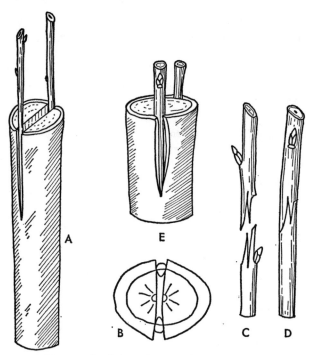

Fig. 30   A & B cleft grafting, C & D tongue grafting,
E bark (crown) grafting

ultimately union is achieved between stock and scion. After insertion of the scion, the flaps of the bark of the stock are folded over, the whole tied with raffia and waxed over. This increases the chances of contact between the cambia (actively growing regions) and lessens the chances of drying out or of infection by disease. Most budding operations are carried out in the summer and union between the two parts may be achieved in about a fortnight. The buds, however, may not grow until the following spring.

Try to do this with roses. You can either buy stocks from a nurseryman or find a few shoots of a wild rose in hedgerows or on some waste ground. Take a few, remembering that each should have some roots, and grow them on well in your garden. Then cut them down almost to ground level and proceed to bud them

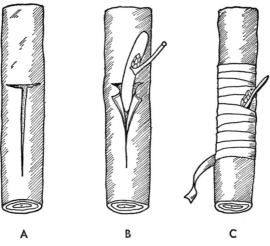

Fig. 31   Budding a rose

as described above and shown in the drawings (Fig 31), having already obtained the buds from any of your favourite rose trees. Insert at least two buds on each stock. Leave the plants for the winter, and in the following spring cut back the stock right to the budded region. This will stimulate the buds to develop. Do not allow any suckers from the stock below the ground to grow. Cut them off below the ground as near to their origin as possible for they only rob the young developing buds of nutriment.

Usually, as has been said, the characters of stock and scion are seen not to be transmitted to one another. There are apparent exceptions; it has been found that leaf mottling in some plants can be transmitted by grafting. The reason for this is that the mottling is due to a virus disease, and when a graft is made transmission of the virus takes place. Virus diseases of potato, tomato, tobacco and many other crop and garden plants are well known and cause very serious losses. The transmission of these has been studied, and tomato and potato are favourable plants for this kind of experiment.

You can graft tomato and potato by cleft grafting. For the scion use a shoot a few inches long and remove most of the leaves except the smallest. Then cut the base of the shoot neatly to a sharp point (Fig. 32). Select a suitable plant, not too young or too woody, for the stock, cut off the main shoot at a point where the diameter of stock and scion is the same and then make a

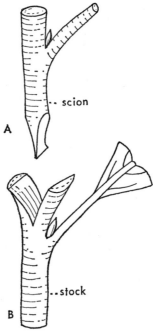

Fig. 32   Cleft grafting for potato and other similar plants

vertical cut in the end. Place the scion in this cut and bind the two together with raffia (previously softened with warm water) so that both are firmly held together. A good material to use for this and incidentally for all grafting is the rubber tape found in the middle of golf balls. By virtue of its elasticity it holds stock and scion together and the loose end can be sealed with a little rubber solution. Keep the grafted plant away from the direct sun for a few days. In the case of tomatoes and potatoes the union will take up to about a fortnight to complete.

If you have a tomato or potato plant with leaf mottling, or a potato with leaf roll or some other suspected virus disease, you can try transmitting it in this way. If similar symptoms are transmitted you can be sure you are dealing with a virus infection.

Another, perhaps easier, way of grafting such plants as tomatoes, cucumbers and other plants which have long flexible stems is called inarching. A notch is made in each stem, one pointing up-

Fig. 33   Inarch grafting

wards and the other downwards. One is then slipped into the other (Fig. 33) and the graft tied up as before. When the union is complete, one or other of the stems may be cut so that either plant can be made the scion or stock as desired.

Very curiously, a plant may contain a virus disease without any visible effects. The popular variety of potato 'King Edward' is a case in point, for it is infected with a virus known as paracrinkle. If 'King Edward' is grafted on to the variety 'Arran Victory' or other susceptible variety, then a severe distorting disease results. If you happen to have some 'King Edward' potatoes you can try growing them and graft them by inarching on to potatoes of other varieties and seeing what happens.

There is another and very ingenious way of grafting potatoes using the tubers. All you need is some cork borers or even an

D

apple corer will do. Take two varieties of similar size, say one of 'Arran Victory' and one of 'King Edward'. Take two cork borers, one slightly larger than the other, and with the smaller of the two cut a core from the 'Arran Victory', being careful to include an eye. With the larger of the two borers cut a similar core from the 'King Edward' and insert this core into the hole in the tuber of the 'Arran Victory' where it will fit tightly. Then grow both the implanted tuber and the core cut from the 'Arran Victory'. Compare the two as they grow; the 'Arran Victory' core should grow into a normal healthy plant to serve as a comparison with the other.

16 Soil bacteria

SOIL is anything but the inert material it appears to be. It is teeming with living organisms, most of them so minute that they can only be seen with the high power of a microscope. Even the large organisms like earthworms, eel worms, insect larvae are far more numerous than you would think for an acre of field may contain as many or more than 50,000 worms alone. The smallest organisms are the bacteria, all-important because they bring about the processes of decay whereby the dead remains of organisms are broken down to their original components to be used afresh by living organisms. One can show the presence of living organisms in soil by observing the changes they bring about. Milk is turned sour by bacteria, and this can be made the basis of a simple experiment.

Take a small amount of garden soil, remove from it all stones, twigs and the like and divide it into two halves. Put a spoonful or two into a small tray and bake it in an oven. Then take two conical flasks from your chemical set and boil water in them for ten minutes or so. Boil also some milk in a saucepan, empty the flasks of their boiling water, add immediately a little of the boiled milk to each and plug both the flasks at once with a light plug of fresh clean cotton wool. Carefully remove one of the cotton-wool plugs, holding it in your hand all the time, shake a little of the baked soil into the flask, and then re-plug. In the same way add a similar quantity of unbaked soil to the other flask and re-plug. Keep both in a warm place for a day or so and then remove the plugs and smell the milk. One will have gone bad and will smell accordingly, while the milk in the other flask will remain fresh (Fig. 34).

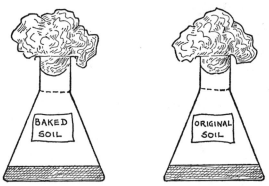

Fig. 34   Baked and unbaked soil added to milk

Some of the most important and interesting bacteria are those concerned with nitrogen fixation, i.e. the uptake of nitrogen from the atmosphere and its incorporation in the body of the bacterium. There are really two categories, those that live in association with other plants and those that are free-living. Dig up a clover or medick plant from a field, or, if you prefer, a pea or bean from your garden. Look at the roots carefully and you will find they bear small swellings or nodules (Fig. 35). These nodules contain bacteria which are able to absorb and use the nitrogen of the air that is present in the soil. This gas is passed from the bacteria to the parent plant as a nitrogen compound which the plant can use. So the Leguminosae, the family to which these plants belong, has an advantage over the rest of the plant world in this respect, and many can flourish where the combined nitrogen content of the soil is very low. This association is also one of the reasons for growing clovers, lucern and other legumes as fodder crops, for if you plough in the roots, the soil is enriched in nitrogen, quite apart from the yield of the crop itself.

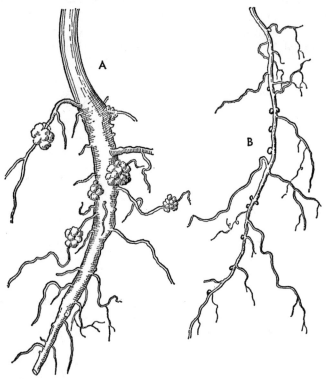

Fig. 35   Roots of lupin and clover showing bacterial
nodules

Quite simple experiments can show that the root nodules do not form in soil in which all organisms have been killed. Take a quantity of soil sufficient to fill a five or six inch pot and sterilise it by boiling it in water for ten minutes in an open saucepan. Then gently boil the excess water away until it is dry enough to put in a really clean pot. Fill another similar pot with ordinary soil and sow equal numbers of runner beans or peas in each. Allow them to grow till they are about a foot high (one-third of a metre) and then tip the plants out of the pots, and wash away the soil from the roots. Only the plants grown in ordinary soil will show the presence of nodules. If you have a microscope you can try making a preparation (Chapter 17) of one of the nodules to see the bacteria, though these are very small indeed.

There are other bacteria that fix nitrogen in the soil as well as those which do it in association with the Leguminosae. One of them is *Azotobacter* and it can be made the basis of some interesting experiments first described in a Pelican book published a long time ago.*

Azotobacter is a bacterium which needs the presence of oxygen and will grow on the surface of soil, provided the nutrients it requires are present, and provided the surface does not dry up. If conditions are really favourable the bacteria grow and divide to form a tiny pearly rounded colony large enough to be seen with the naked eye. Other organisms, such as moulds, may grow as well, but they are quite different in appearance. You can count the colonies, and on a suitable surface they take a week or so to appear.

To carry out a simple experiment you require the following chemicals: potassium phosphate (dihydrogen phosphate), acid potassium phosphate, some mannitol and some precipitated chalk. You may have the phosphates in a chemical set, but they are reasonably cheap and easily obtained substances. If mannitol cannot be obtained, starch can be used instead. Rice starch is best because it has the smallest grains and all the chemicals should be finely powdered. You will require a number of small pots – paste pots or ointment pots are excellent for the purpose. They should have a diameter of about one inch or so (3–4 cm) and to make the best comparisons they should all be the same size.

The two phosphates supply potassium and phosphorus which are both essentials for bacterial growth. One, however, is rather acid, the other alkaline, but since *Azotobacter* prefers neutral or slightly alkaline conditions it is best to make a mixture of the

---

* *Microbes by the Million*, H. Nicol, Penguin, 1939.

two, which is neutral. Therefore mix thoroughly two parts by weight of potassium phosphate with one part by weight of potassium acid phosphate. Dissolve one part of this in 1000 parts of distilled water, say one gram in a litre of water. Shake well to get the salts to dissolve. This should be neutral but you can test it with the B.D.H. indicator mentioned in Chapter 2. If on testing it is not neutral, you can add a solution of one or other of the phosphates until it is. If you cannot get distilled water use clear filtered rain water. Tap water may be used, but it is usually alkaline, and so a larger proportion of the acid phosphate will be required to achieve neutrality.

Take a quantity of good garden soil sufficient to fill the number of pots you have decided to use, and spread it out to dry in the air. Good garden soil is most likely to contain *Azotobacter*, but nearly all soil, barring the most acid, does. Then rub the soil through a sieve to remove the stones, roots, etc., and divide it into two halves. To one half add about one per cent of mannitol by weight, or if you are using starch about five times as much. Now mix a little of each with sufficient rain or distilled water to make a smooth paste and fill a pot with each. Smooth off the top with a flat knife blade as cleanly as possible. Now repeat this with the phosphate solution. This gives four pots; repeat the whole with soil to which you have added a little chalk. You will then have eight mud pies as follows:

No mannitol	With mannitol
1. Soil only	soil only
2. Soil+phosphate	Soil+phosphate
3. Soil+chalk	Soil+chalk
4. Soil+chalk+phosphate	Soil+chalk+phosphate

It is good, if you can, to duplicate all the pots for then your results will be the more reliable.

Having set up the pots of moist soil, put them in a moderately warm place, but not so warm as to make the surfaces dry up. In fact, cover the surfaces so as prevent this, but do not allow the cover to touch the surfaces of the soil. This is perhaps the only difficulty with this experiment, for the soil in each of the pots should be uniformly wetted. If it dries out no colonies of *Azotobacter* will appear. But assuming you surmount this difficulty, count the number of colonies on each surface with the aid of a hand lens, and compare them carefully to see if the addition of mannitol, or chalk, or phosphate produce any significant increases. Table 11 shows the result for such an experiment.

Table 11   The numbers of Azotobacter colonies on a garden soil

Treatment	No. of colonies	Average	Increase or decrease
None (control)	15	15	
Chalk only	33	33	+18
Phosphate only	71	71	+56
Chalk and phosphate	122	122	+107
mannitol alone	141, 150, 149, 160	150	—
mannitol and chalk	166, 162, 253, 180	190	+40
mannitol and phosphate	350, 470, 425, 175	355	+205
mannitol and phosphate and chalk	359, 276, 225, 376	309	+159

You may notice that this result, which is not a particularly good one, shows wide variations in the individual totals. In view of this, how far do you think the differences in the last column have any real meaning? Could not one further result alter the plus figures by quite a large amount? Why then is it necessary to repeat (or replicate, to use the right word) the experiment several times?

As has been pointed out, nitrogen compounds from complex substances like proteins to simple ones like amino-acids are broken down by bacteria through many stages until they form nitrates in the soil. The latter stages of these processes, namely the changes from ammonia to nitrate, are known as nitrification and these can be studied by the following experiment. Take a wide glass tube about a foot or so long, preferably narrowed at the base. Fit a glass wool plug into the narrowed part, or if the tube is not narrowed tie a piece of muslin round the base. Fill the tube almost to the top with some sandy loam soil without roots or stems or other large particles and shake it down so that it is evenly and closely packed. The glass wool or muslin is to keep the soil from falling out (Fig. 36). Next, support it vertically and

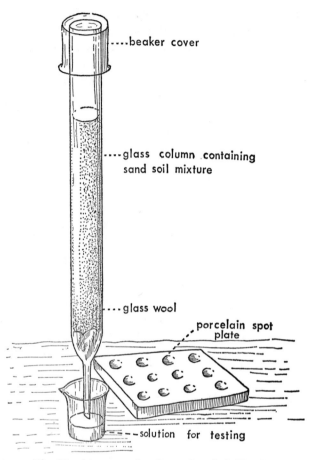

Fig. 36   Apparatus for the study of nitrification

prepare a solution of 0·2 per cent ammonium sulphate and slowly pour 50 ml of this over the soil. Collect the solution that runs through and test it for ammonia, nitrite and nitrate as follows. Put a drop of the solution on a tile, using a glass rod or tiny pipette, and add a drop of Nessler's solution. A yellow or orange colour indicates the presence of ammonium compounds. To a second drop add a drop of sulphanilic acid reagent when a red colour after a few minutes indicates the presence of a nitrite. Thirdly a blue colour when a drop is mixed with a drop of diphenylamine reagent (CARE) indicates the presence of a nitrate.

Continue to add 50 ml portions of the ammonium sulphate until no reaction for nitrate or nitrite is obtained. Then the soil contains only ammonium compounds. Cover the top of the tube so as to prevent drying up, and if necessary add a little distilled water to keep the soil moist. Leave the column for a week, and after this, run sufficient distilled water through it to collect a little at the outlet at the bottom. Test this as before, when you should find that some nitrate has been formed.

17 The use of the microscope

MICROSCOPES are easier to come by today than at any time in the past and for a modest outlay of a few pounds quite a reasonable instrument can be obtained. Since it may be anything from a comparatively simplified modern instrument to something old and complicated, it is only possible to give some rather general notes on how to use it. An ordinary pocket lens gives you a magnified image and extremely useful it is for much botanical observation. A microscope, or more correctly a compound light microscope, gives you a double magnification. There is a lens at the base called the objective which produces a magnified image which is then further magnified by a second lens, the eyepiece. These lenses are mounted in a tube fastened to the body of the instrument in such a way that it can be moved up and down for focussing. The object to be viewed is usually placed on a glass slide on the stage of the instrument. The microscope views all objects by transmitted light, so microscope preparations must be transparent. This is most important and often forgotten by the beginner who tends to have everything present in too great a thickness to see anything well. The microscope has a mirror to reflect light through the instrument so that the object is clearly illuminated.

It is possible to use a microscope to look at almost anything, but this short chapter is mainly concerned with plant life in water. Visit a small pond, or a long-standing puddle, and collect some green pond scum, as it is usually called. Using a pocket lens you can see that it consists of fine green threads. Place a very few of these in a drop of water on a microscope slide, cutting them, if

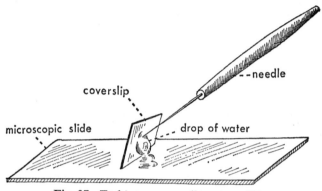

Fig. 37   To lower a coverslip onto a slide

necessary, with a sharp penknife. Take a cover glass, touch the drop of water with one side of it while supporting the other side with a mounted needle (Fig. 37). Then lower the cover glass gently on to the slide and your preparation is ready. The drop of water should be just sufficient to form a film between the cover glass and the slide. Any surplus should be carefully removed with blotting paper and there must never be any water on top of the cover glass.

Next it is necessary to set the instrument for use. First of all adjust the mirror so that light is reflected up the instrument. You can do this by removing the eyepiece, looking down the tube and moving the mirror until maximum illumination is obtained. You will find that this does not necessarily give you the maximum definition when you examine the object, but the light can later be reduced to do this by means of an iris diaphragm under the microscope stage. Now replace the eyepiece, place the slide on the stage, and turn the nosepiece of the instrument so that you are using the lowest power. (This is the objective with the widest lens.) Now with the coarse adjustment of the instrument move the objective downwards until it nearly touches the slide. Then with your eye in position focus carefully by racking the lens upwards. After a little practice, you will get used to the instrument and you will be able to do this quickly.

You will notice the fine threads of the plant to be made of boxes called cells, and they may look something like on the drawings in Fig. 38. *Cladophora* is a very common pond scum, usually rather yellow-green in appearance and rough to the touch. *Spirogyra* is a bright dark green and smooth and silky to the touch, and there are many others.

Now turn to the high power to see more detail of this tiny green plant. The finer points of the structure can be seen better if the plant is suitably stained. Iodine is a very easy stain to use and this turns the nucleus of the cell a dark brown and any starch grains present a bluish black. Place a drop of the iodine on the edge of the cover slip and draw it through with a piece of blotting paper (Fig. 39).

There are much smaller forms of plant life than the visible threads of pond scums. The green powder you can see on tree trunks is a tiny one-celled organism called *Pleurococcus* which can be easily mounted and examined. There are many others to be seen in pond and river water and some of them are free-swimming. All of these green microscopic forms of life belong to one of the lowest groups of the plant kingdom, the green algae. Some of them are macroscopic, like the bright green sea lettuce of our sea shores, and the tubular green growths found in rock pools near the sea known as *Enteromorpha*.

You will also want to look at some parts of the flowering plants with the microscope. One of the easiest things to see are the cells from the pith of a berry and almost any one will do. Scrape a very tiny quantity of the pith on to a slide in a drop of water and flatten with a cover slip. It is also useful to stain with iodine. These cells should show the outer envelope or cell wall, the nucleus,

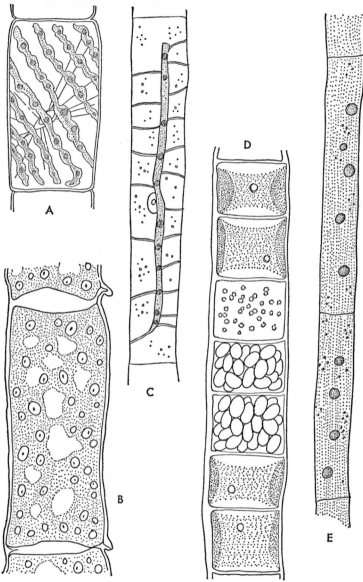

Fig. 38 Various kinds of Algae found in freshwater:
A, Spirogyra; B, Cladophora glomerata; C, Mougeotia
capucina; D, Ulothrix aequalis; E, Mougeotia viridis

stained brown by iodine, occasional chloroplasts (green bodies) and starch grains stained bluish black with iodine. The whole of the central part of the cell which looks empty is full of cell sap, while the wall is lined on the inside by a layer of living material, the cytoplasm, in which the nucleus is embedded (Fig. 40).

Other things to examine with a microscope include the leaves of the Canadian pondweed (*Elodea canadensis*). Take a young leaf of this water plant and mount it face downwards in a drop of water. Then find the mid-rib of longer cells and focus very carefully

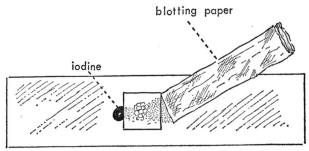

Fig. 39   Irrigating a preparation with iodine

on the chloroplasts in them with the high power. They will be seen to be slowly moving round the cell, carried by the cytoplasm. Try to see if the direction of circulation is the same in all cells, and see if you can measure the time a chloroplast takes to travel round one cell. Protoplasmic movement is not confined to the cells; it can be seen in many other plants (Fig. 41).

Easy microscopic objects to examine include the hairs on the leaves and stems of plants for these can easily be scraped off with a knife or needle and mounted as before. Here again a little staining is helpful. Plant hairs may be simple, one-celled un-branched structures, or multicellular outgrowths looking like miniature trees; they may be glandular with cells at the head that secrete substances of various kinds, or even more complex like the stinging hairs of a nettle (Fig. 42).

Pollen grains are also easy things to examine with a microscope. Just shake or rub a little of the pollen from the stamen of a flower

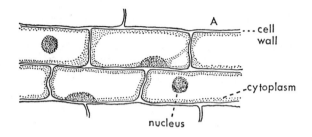

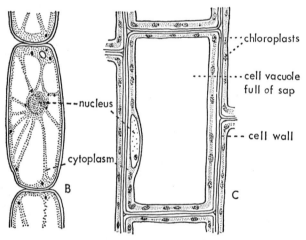

Fig. 40 Three examples of plant cells: A, Onion; B, Spiderwort; C, Typical leaf cell

on to a microscope slide and examine them first as dry objects, then mounted in water, and then mounted in a stain (e.g. safranin) dissolved in methylated spirit. You will see their varied shape, their difference in size and the differing ornamentation. Pollen grains have an outer wall called the extine which is variously sculptured and an inner wall or intine which is thinner. At maturity the pollen grain contains two nuclei, but they are difficult to see owing to the fact that the walls of the grain stain so well that the contents cannot be seen. Occasionally, you can see them as in the spiderwort (*Tradescantia virginiana*), where one nucleus is long and narrow and the other small and spherical, especially if a nuclear stain like acetic methyl green is used (Fig. 43).

You can also try to germinate some pollen grains in sugar solution. The majority of pollen grains germinate in a sugar

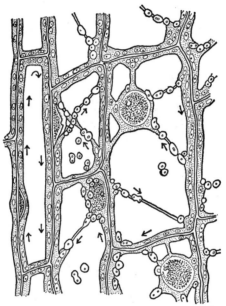

Fig. 41   Protoplasmic circulation in *Elodea*

solution if it is of the right strength. Though this is rather a matter of trial and error the following concentrations are suggested:

> Peony 5% sugar
> Spiderwort 5–8% sugar
> Sweet pea, other peas, and the tulip 15% sugar
> Onion (*Allium* s.p.p.) 5% sugar
> Busy Lizzie (*Impatiens sultani*) 5% sugar

In each case the sugar solution may be set by the addition of one and a half per cent gelatine. A drop of this mixture is put on a

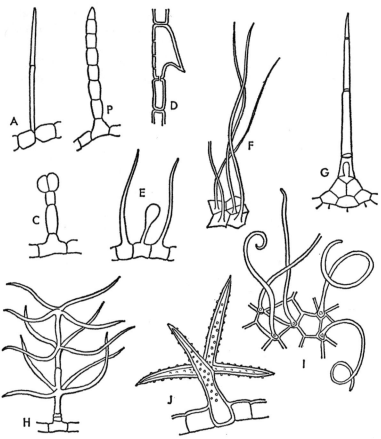

Fig. 42 Different kinds of plant hairs—from:

A. corolla of *Epigaea* F. stem of *Onopordum*
B. leaf of *Coreopsis* G. leaf of *Cucumis*
C. corolla of *Phryma* H. young leaf of *Platanus*
D. leaf of *Avena* I. fruit of *Rubus*
E. calyx of *Heliotropium* J. stem of *Aubretia*

slide, and the grain shaken on to and a cover slip placed in
position. Better still, a hanging drop culture may be set up as
this allows of a better supply of oxygen to the grains (Fig. 44).
This is set up by using a small plastic ring which is stuck on to a

slide with vaseline. A thin drop of the gelatine sugar mixture is placed on a cover slip, allowed to set and the pollen grain sprinkled on to it. This slip is now sealed to the ring face downwards using a little vaseline. Care should be taken to make the

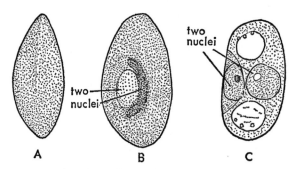

Fig. 43   Pollen grains of spiderwort
A, dry; B, in water showing two nuclei; C, young grain
in water showing the two nuclei at an earlier stage

drop a thin one, otherwise it will not be possible to focus sufficiently far down with the high power to see them.

Another way to grow pollen is to place the sugar solution in a saucer, and on the top of this float a small square (about 1 cm)

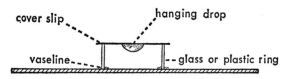

Fig. 44   A hanging drop culture

of thin cellophane, on the top of which the pollen is scattered. Another saucer is then placed on top to keep the whole humid, and after a few hours or longer, the cellophane squares are transferred to a microscope slide, a cover slip placed in position and the preparations examined.

Protoplasmic circulation can often be seen in the tubes produced by pollen grains, often very beautifully, and there are also

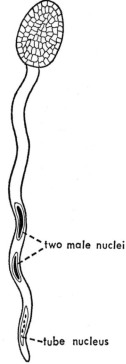

Fig. 45   Germinating pollen grain

three nuclei, for one of the two in the mature grain divides; these two nuclei are incidentally the male cells or gametes of the flowering plant (Fig. 45). Most pollen grains develop within hours of being sown on the right sugar solution and some grow very quickly indeed. The greenhouse plant *Gloxinia* (*Sinningia*) *speciosa* has grains which germinate in ten per cent sugar solution in about two hours, and are said to grow so rapidly (several mm per hour) that with strong magnification, the tip can be seen to move across the field of view, provided, of course, that you have enough patience to glue your eye to the microscope for some minutes.

18 Mosses and liverworts

SMALL green plants growing on the ground, on tree-trunks and on rocks mostly belong to the group of plants known as Bryophyta. This includes two distinct sub-groups, the mosses and liverworts. You may be surprised to know that there are some 600 species of moss and over 200 species of liverwort known in the British Isles. They should not be confused with lichens which grow in similar places to mosses but which are grey-green, yellow or black, but never a really bright green. You can easily make your own collection of mosses. Just allow them to dry, and they will keep indefinitely packed away in small envelopes. When you want to look at them soak them for a while in water, when they will swell and appear as fresh as when they were first gathered. Mosses make excellent subjects for study with the microscope, and in practice mosses can be almost completely identified from the microscopic characters of their leaves. This is beyond the scope of this book, but practise mounting a single moss leaf quite flat in a drop of dilute glycerine under a cover slip. It is not all that easy, but in good preparations you can see the whole size and proportions of the leaf, the size and shape of the individual cells, the character of the margin, the pattern of the mid-rib if present and many other features (Fig. 46).

It is also quite easy to grow many mosses, to make a moss garden, in fact. You must, of course, try to imitate the original circumstances in which you find the moss. For example, most like a moist atmosphere and don't like bright sunshine. Many mosses will grow in shallow trays with a little soil on which you can put a layer of stones, or small fragments of rocks, or peat or sand, or rotten wood, according to the species you wish to grow. The mosses should just be pressed into this but they should have a little of their original soil adhering to them. They should be watered frequently and covered with a glass plate. This prevents the mosses drying out in hot weather. For the chalk mosses you must have a chalky type soil, and equally for the bog mosses one must have an acid peat and water it with rain water so as to avoid adding any lime. The tray of moss should be kept in a place sheltered from the wind and also from too high temperatures, but in this you must be guided by experience. Some mosses

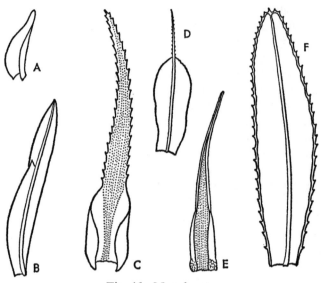

Fig. 46   Moss leaves

A. *Andreaea petrophela*   D. *Tortula intermedia*
B. *Fissidens polyphyllus*   E. *Campylopus schwarzii*
C. *Polytrichum alpinum*   F. *Mnium affine*

will grow much better than others. You will find it necessary to remove ordinary plants as 'weeds' and it will probably also be necessary to remove some of the mosses which grow too fast to give the more choice mosses you wish to grow a better chance.

Some of your mosses will almost certainly form capsules which release many tiny spores which ultimately grow into new moss plants. These capsules have elaborate structures which aid and regulate the dispersal of the spores. Each capsule usually has a hood-like cap which drops off to reveal the lid of the capsule.

When the lid is cast off, there is left a ring (sometimes double) of what are called peristome teeth. These peristome teeth are wonderful objects to see with a microscope, and the variation in them to be seen in different species is quite amazing. Find a capsule and look at it as a dry object on a slide; then cut the peristome off the top of the moss capsule with a sharp razor blade, and again look at it with the microscope as a dry object. The actual detail will depend on the particular moss you have picked upon, but some examples of moss peristomes are shown in Fig. 47.

The teeth of the peristome are very sensitive to slight changes in the humidity of the air. These movements can easily be seen by breathing on a capsule and observing the peristome with a lens or the low power of the microscope. In some peristomes the teeth bend inwards in moist air and outwards in dry, in others inwards in dry and outwards in moist air. The outward bending allows the spores to escape; but in some peristomes there is a second insensitive set of teeth which act as a sieve and allow the spores to escape gradually. It is thought that in those mosses where the peristome opens in dry air, dry conditions favour the dispersal of the spores, and where the teeth open in moist air damper conditions are required, though I do not think this has been clearly proved.

You can make use of your ability to grow mosses to investigate the spore dispersal. Take a long seed tray or box, and fill it with sterile J.I. or other compost. Press the surface down with a block of wood so as to make it smooth. Plant a row of moss cushions along one of the short sides of the box, using a species that is near to ripening its capsules. Tilt the box slightly as in the diagram (Fig. 48). When the capsules are ripe produce a draught of air with a hair dryer (in the cool position) or an electric fan for a few minutes. Keep the soil surface moist and covered as for a moss garden, and in due course you may see small plants springing up. Before this happens you may see a kind of green colour on the surface of the soil. This is called the moss protonema, for a moss does not grow directly to a new plant, but first produces a green thread-like growth on which buds are formed which then

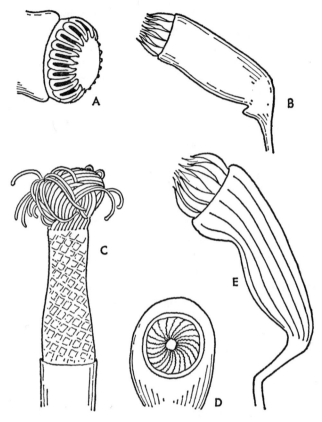

Fig. 47   Peristomes of mosses

A. *Catharinea undulata*   D. *Amblydon dealbata*
B. *Cynodontium virens*   E. *Timmia austriaca*
C. *Tortula subulata*

grow into new moss plants. You may be able to take off some of
the threads of the protonema and to examine them with the
microscope.

Alternatively you can germinate the moss spores and obtain
the protonema directly. To do this take a small dish or saucer and
cut several thicknesses of blotting paper to fit it so as to make a

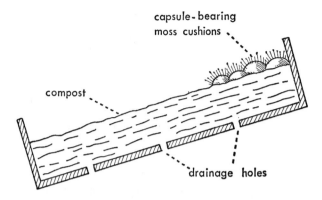

capsule-bearing
moss cushions

compost

drainage holes

Fig. 48   Spore dispersal in mosses

firm pad. Moisten this thoroughly. Now take some ripe capsules,
break them up and scatter the spores on to the blotting paper.

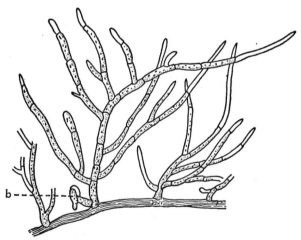

b

Fig. 49   A moss protonema

Cover the dish, and take care to see that it does not dry up. In
a short time the protonema will be seen on the surface of the
blotting paper, and you can examine it with the microscope. It

consists of branched filaments and often some of the cross walls of the cells are oblique; in the axils of the branches some tiny buds develop which grow into new moss plants (Fig. 49).

19 The Fungi

WHEN fungi are mentioned very likely the first examples that come to mind are the common edible mushroom and a woodland toadstool. Actually the toadstool or mushroom is only the fruiting body of the fungus; what might be called the real plant body consists of many fine white threads growing in the soil, and which can often be seen adhering to the base of the toadstool as you pull it from the ground. The toadstool produces literally millions of spores from the flat radiating plates or gills on the underside of the cap. The spores are minute and are carried everywhere by air currents, and so the fungus is dispersed. The spores vary in colour with the kind of fungus and you can make 'spore prints' of them. To do this dip some sheets of white paper separately in melted paraffin wax in a flat dish; then hold them up so that the surplus wax drains away and allow them to dry. Then take a fresh toadstool, cut or break away the cap from the stalk and place it face downwards on the sheet of waxed paper for two or three days. The spores in their millions will drop from the gills and form a pattern of radiating lines on the paper which is a reflection of the pattern of the gills themselves. If you *gently* warm the paper, the wax will melt just sufficiently to stick the spores in position, and you will have a permanent record of this particular fungus. Since the colour of the spores is different for many of these fungi, you can make a collection of spore prints. One thing you will find is that the colour of the spores is not necessarily the colour of the gills.

You can try making similar prints with other plants that produce spores, for example, ferns. The fertile fronds of ferns bear brownish sori on their undersides, and these sori contain many sporangia which are tiny spherical boxes that contain many spores. The sporangia can be seen with a pocket lens. Each of these sporangia springs open to liberate the spores, which though larger than the spores of fungi, are still produced in vast numbers. Spore prints can be made for these in the same way as for the fungi, but the spores of ferns are all of a similar brown colour, and the interest of the spore prints lies in the fact that they show different patterns reflecting the shape and arrangement of the sori on the underside of the leaf.

Mushrooms and toadstools mostly live on the dead remains of plants and animals and are consequently known as saprophytes. There are many other smaller fungi which live in the same way, for example, the moulds which are found so frequently on foodstuffs like jam and cheese, as well as on dead leaves and fruits and other material undergoing decay. Any one of these moulds can be examined with the microscope. The blue-green mould so common on rotten oranges and rotting foodstuffs is a *Penicillium*, though it is not the kind from which penicillin is obtained, which is a much rarer species.

It is fairly easy to grow some of the aquatic fungi which are saprophytic. To do this some food material is required, and for this you can use almost any kind, but perhaps it is as well to try to start with something like a dead fly, or a small piece of apple, or a soaked seed or a piece of egg white. Whatever you decide to use, tie it to a piece of cotton and suspend it in the middle of a large jam jar full of fresh pond or river water. After a few days a fine white mould will be seen growing from the bait. This mould looks like thin cotton wool and some of it should be transferred as gently as possible to a microscope slide and examined with the microscope. Most of the aquatic fungi obtained in this way are species of *Saprolegnia, Achyla, Pythium* and you can see the fine tubes (called hyphae) with the high power of the microscope. Many of the aquatic fungi reproduce by setting free very tiny little spores that are able to swim about in the water and which ultimately give rise to new plants. You may be lucky to see these in the water, but if not, try applying a little pressure to the cover slip. This may be just enough to release some of them from one or two of the hyphae so that you can see them.

The culture and collection of aquatic fungi is quite easy and you can make special traps for them. Take a piece of perforated zinc and fold it round a broom handle so as to make a cylinder about 10 cm long. Then wire the folded edges together and fit a cork to both ends. If you can get the cylinder ready-made so much the better. Make several of these and place a bait in each. The bait can be anything of your own choosing, but hemp seed is good, as are also small fruits like grapes or rosehips. Tie a string to the trap and lower it into a pond or stream. After some days or weeks haul out the trap and examine the bait for aquatic fungi as you have done previously (Fig. 50).

You will undoubtedly find other organisms amongst the fungi growing on or near the baits you have provided. There will almost certainly be some small animals belonging to the protozoa and ciliata, and there will be millions of minute bacteria.

Saprolegnia does not always live on dead materials; occasionally it attacks the gills of fish causing what is called 'salmon disease'. It also attacks the body surface of the fish if it has been damaged by some of the protective scales having been knocked off. Here the fungus is living as a parasite causing a disease; numerous other diseases, particularly of plants, are caused by fungi.

One such disease is caused by a fungus related to *Saprolegnia* called *Pythium* and it causes the 'damping-off disease' of seedlings (Fig. 51). If seedlings like cress, tomato or snapdragon are sown close together, kept covered with a glass cover so that the internal atmosphere is moist and warm, then they develop a brownish region just above soil level, topple over and die. Some of these infected parts should be mounted on a microscope slide and crushed by a little pressure on the cover slip. The fungus can be seen growing between the cells of the seedling, particularly if a little stain like cotton blue is added to the preparation.

One of the most important fungus diseases is potato blight. It is a fungus which has changed the course of history, for it caused the Irish famine in 1844–5 and thus led to the repeal of the corn

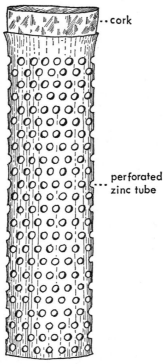

Fig. 50   A fungal trap

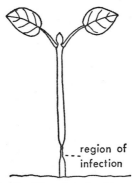

Fig. 51   Seedling showing 'damping off'

laws. The disease is found in potato plants in July where it forms brown patches on the leaves. The fungus spreads from plant to plant by airborne spores which are produced on branched

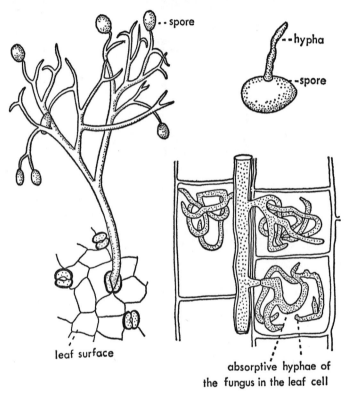

Fig. 52   Phytophthora (Potato blight)

structures looking like miniature trees under the microscope (Fig. 52). These are produced from the underside of the leaves, and give the diseased parts a silky appearance. To obtain plenty of them for examination find some infected plants by searching a potato crop and place some diseased leaves under a glass jar which is wet on the inside so the atmosphere is very moist. Leave the leaves for about two days when they will be seen to be covered, especially on the underside, with a white mould formed by the branched spore-bearing structures. These are particularly well developed at the edges of the brown spots. Try placing a small piece of the leaf on a microscope slide and examining it as a dry object. Some of the minute lemon-shaped spores may be seen on their stalks, though a great many will have been knocked off in making the preparation. Try also mounting a tiny part in water, when the branched threads which support the spores should be seen.

Careful examination of many wild and garden plants will show you numerous other examples of plant diseases, which you can examine in a similar way with your microscope. Mildews are often seen as white moulds on leaves, while the brown streaks on the leaves of grasses and cereals are known as rusts. In diseases known as bunts, the ovaries are transformed into masses of black or dark coloured spores. Try looking for these on some of the common wild grasses like the false oat or the cultivated cereals like barley.

Appendix

The laboratory chemical mentioned in this book can mostly be supplied by a local chemist, but in cases of difficulty, larger firms will usually supply small quantities by post, e.g.:

> Messrs B.D.H. Ltd,
> West Quay,
> Poole,
> Dorset.

Industrial methylated spirit cannot be bought free of duty in Great Britain without a licence. Iso propyl alcohol can be obtained (see above) and is an effective substitute.

All the materials required for composts, including the chemicals, can be bought from a local horticultural supplier.

Slides, coverslips and stains for use with the microscope can be bought from specialist firms, e.g.:

> E. Gurr Ltd,
> 42 Upper Richmond Road,
> East Sheen,
> S.W.14.

Tomato plants for genetic experiments can be obtained from

> P.P.G.,
> 18 Harsfold Road,
> Rustington,
> Sussex

and irradiated seeds from

> Carolina Biological Supply Co,
> Burlington,
> N. Carolina 27215,
> U.S.A.

Index